Aid Performance and Climate Change

The richer countries spend about US$165 billion yearly on overseas aid, mainly to keep human development going. These efforts are undermined by climate change, water-catchment damage, biodiversity loss, and desertification, and their interactions with social systems at all scales, which few aid designs or evaluations fully address. This must change if aid performance is to be improved.

Constraints to be overcome include limited understanding of the very complex systems that aid investments affect, and of the ecology behind climate change adaptation and mitigation. *Aid Performance and Climate Change* targets these problems and others, by explaining how to use multiple points of view to describe each aid investment as a complex system in its own unique context. With examples throughout, it reviews cases, ideas, and options for mitigation using technology and ecology, and for adaptation by preserving resilience and diversity, while exploring related priorities, treaties, and opportunities.

Combining an empirical, eye-witness approach with methodological conclusions, this book is an essential resource for those looking to improve aid design and evaluation, and will be a necessary tool in training the next generation of aid professionals to respond to the causes and consequences of climate change.

Julian Caldecott has a background in wildlife management and rainforest conservation. Since 2000 he has evaluated major aid investment programmes for the EC, UK, Norway, Finland, and Switzerland, five donors that together contribute more than a quarter of all official aid. His work focuses on design and performance issues concerning climate change, biodiversity, ecosystem management, and related matters of sustainability and institutional and community development.

"In over 30 years of working in climate, environment and international development this is the most extensive analysis of the lessons from climate and environment interventions based on solid evaluations that I have come across. A must-have for sustainability practitioners and evaluators, and a book I wish I had been able to read many years ago." – *John Carstensen, Head of Profession, Climate and Environment, Department for International Development (DFID), UK*

"This book uniquely captures the skills of a good evaluator as it intelligently combines what needs to be assessed and how to present evidence from a variety of diverse projects and programmes. It offers an essential guide to anyone interested in project/programme design as well as learning how to evaluate. It is the go-to guide that every consultant, donor, and evaluation manager must read. For those starting or learning about project/programme management, look no further than this book – it's pure bottled experience." – *Karen Lawrence, Impact Measurement Manager, Zoological Society of London (ZSL), UK*

"This book is a hugely valuable contribution to the professionalisation of aid evaluation; a challenge to the prevalent 'conspiracy of optimism' in designing and delivering aid investments; a fascinating source of knowledge on the various kinds of investment aimed at climate change mitigation and adaptation; and a smart guide for navigating the contextual challenges associated with evaluation practice." – *Letisha Lunin, Monitoring and Evaluation Coordinator, Climate and Development Knowledge Network (CDKN), PricewaterhouseCoopers LLP, UK*

"The evaluation methods described in this book have several strengths. Scoring requires both evidence and analysis, and the process has to be documented in a transparent and systematic manner. As a result, the scores encourage a comparative approach and can promote dialogue. The methods are thus powerful tools, and will be very useful if evaluators are well trained, and if those evaluated are not too defensive to let them be used." – *Carmenza Robledo, ECOEXISTENCE – Robledo Abad Althaus, Switzerland*

Aid Performance and Climate Change

Julian Caldecott

Routledge
Taylor & Francis Group
LONDON AND NEW YORK

earthscan
from Routledge

First published 2017
by Routledge
2 Park Square, Milton Park, Abingdon, Oxon OX14 4RN

and by Routledge
711 Third Avenue, New York, NY 10017

Routledge is an imprint of the Taylor & Francis Group, an informa business

British Library Cataloguing-in-Publication Data
A catalogue record for this book is available from the British Library

Library of Congress Cataloging-in-Publication Data
Names: Caldecott, Julian Oliver, author.
Title: Aid performance and climate change / Julian Caldecott.
Description: Abingdon, Oxon ; New York : Routledge, 2017.
Identifiers: LCCN 2016046625| ISBN 978-1-138-29446-2 (hb) |
ISBN 978-1-138-29448-6 (pb) | ISBN 978-1-315-23149-5 (ebk)
Subjects: LCSH: Economic development—Environmental aspects. | Economic assistance—Evaluation. | Global environmental change—Prevention. | Climatic changes—Prevention.
Classification: LCC HD75.6 .C297 2017 | DDC 338.91—dc23
LC record available at https://lccn.loc.gov/2016046625

ISBN: 978-1-138-29446-2 (hbk)
ISBN: 978-1-138-29448-6 (pbk)
ISBN: 978-1-315-23149-5 (ebk)

Typeset in Goudy
by FiSH Books Ltd, Enfield

Contents

Tables and boxes

Tables

Boxes

Preface

This book draws on my 40 years' work in the tropics as an ecologist and consultant, and especially the last 15 when I mainly did aid evaluation. This was when climate change moved into the heart of the aid world, as human priorities began to adjust to ecological reality. Fifteen years is an instant in geological time, but in human terms it was when the threat of climate change was recognised, its links to mass extinction and the spread of deserts appreciated, and the first real steps were taken to do something about it. The response was at first more focused on reducing net GHG emissions than in helping developing countries adapt to the consequences of historical emissions. In the 2000s, Norway, several EU Member States, the EU and other OECD members and international agencies were actively investing or preparing to invest in ecological mitigation (REDD+, etc.) and increasingly in technological mitigation (renewable energy, etc.), but the more complex resilience and diversity agenda of adaptation lagged behind. Thus the Rio Climate Marker on mitigation was agreed in 1998, but on adaptation only in 2009. By 2007, however, the main contours of the climate change challenge and response were clear. Aid professionals were in a good position to see the big picture, so it was in the aid world that experimentation was most vigorous. These efforts had to push against a lack of experience, incomplete consensus, and patchy political support, but much was achieved even so. Many lessons have been learned, and this knowledge can guide the investments that are needed if we are to survive beyond the next few decades as a prosperous species in a functioning biosphere.

Aid investments now add up to about US$165 billion annually, and adaptation is an increasing priority because the climate is demonstrably changing. This is a small share of a US$75 trillion global economy, but it can make a difference if spent well. This book aims to gather together some of the most useful knowledge about how to do that, and also floats some ideas for ramping up change to a more realistic level. We are all now exposed to serious risk, including a real chance of social, political, and economic turbulence at unprecedented levels, possibly suddenly and soon. While there is still a chance of an orderly transition to sustainability, I propose two targets for the aid profession over the next 15 years: (a) to raise the performance of the world's aid portfolio, using the core criteria described here, from an average score of 3.5 ('weak/moderate') to 5.5 ('strong/very strong') out of seven; and (b) to raise the share of aid in which climate change is actively considered as a threat and target, in terms of mitigation, adaptation, biodiversity, and desertification, at all stages of the project cycle, to 100%. Meanwhile we will also need to consider two related questions: on whether removing GHGs from the air quickly enough to prevent disaster should be thought of as 'mitigation aid', rather than simply a cost of doing business and surviving; and on whether easing the movement of large numbers of people displaced by climate change should be thought of as 'adaptation aid', rather than simply a cost of finding humane solutions to what are now inevitable social and environmental challenges.

While I was writing this book in 2016, the mitigation agenda became compromised in several ways. First, a rumbling debate about whether climate change is caused by people or not has continued, and those who deny a primary human role now dominate national politics in the USA and remain influential elsewhere. Second, adoption of the 2015 Paris Agreement's 2°C ceiling on global warming may be having the perverse effect of discouraging attention to worse scenarios, and the riskiest scenarios even at 2°C, almost as if avoiding runaway, catastrophic climate change is now seen as 'job done'. This is despite dangers that are in fact escalating rapidly – for example from melting sea ice in the Arctic. Third, growing tensions over immigration and inequality have become very influential in the EU, UK, and USA, causing a rise in political polarisation and anti-internationalist sentiment. Related factors include the UK's vote to leave the EU, and an active political debate in the UK and USA about the merits and costs of overseas aid. All this is distracting leaders, voters and aid agencies from climate change priorities of all kinds.

It is easy to be discouraged, but we should take hope from the irreversibility of the adaptation agenda, which will encourage investment in many 'no regrets' measures to safeguard the integrity of the biosphere by focusing on water, ecosystems, and disaster risk. And the mitigation agenda will not close down, since some countries, cities, states, companies, NGOs, and individuals will continue to invest in it regardless of national and international discord. We can also be reassured by changes such as that wind in the UK generated more power than coal in 2016, and that recent mitigation investments have allowed some places to run entirely on renewable energy for a while, including Portugal for four consecutive days in May 2016, and Costa Rica for 110 in June to October 2016. In a few years, maybe coherence will be restored to the overall climate change response, and we will be in a better position to make even more rapid progress on mitigation despite the temporary loss of momentum and the permanent loss of time. In any case, and meanwhile, this book is designed to enable more effective investment in adaptation as well as ecological and technological mitigation, for use whenever and wherever it is needed.

Julian Caldecott
January 2017

Acknowledgements

Many people and institutions in many countries were involved in the projects explored in this book, and are acknowledged in the cited reports. Here I wish to thank those people who were special colleagues in multiple projects, or who helped me with this particular book: Sue Bishop, Ben Caldecott, John Carstensen, Trudiann Dale, Mikko Halonen, Finn Hansen, Mochammad Indrawan, Bernadeta Killian, Karen Lawrence, Anu Lounela, Letisha Lunin, Mary Monro, Nick Ostler, Aira Päivöke, Lydia Richardson, Carmenza Robledo, Berthold Seibert, and three anonymous peer reviewers. Others who shaped my recent thinking are no longer living and are greatly missed: Hugh Blackett, Moyra Caldecott, Stratford Caldecott, Don Grimes, Roger Hammond, and Dragana Vilinac.

The evaluations used here are the responsibility of the authors of the relevant reports as cited in the reference lists, and their use does not reflect or seek to represent any official position of the clients listed in Appendix 1, even if they sponsored or published the reports. This book is based on my interpretation of diverse phenomena, and responsibility for the selection and explanation of cases is mine alone.

Some terms are used here that may be unfamiliar, and a glossary is provided in Appendix 2. Where a word is described as having a meaning 'in ordinary English', this refers to the 2006 edition of the *Concise Oxford English Dictionary*. In referencing, page numbers are given where a direct quotation is used. In quotations, minor corrections that do not alter the sense of the original text have been introduced, and acronyms and abbreviations have been expanded for clarity. Throughout the book, preference is given to cases where I was directly familiar with the details, and in quotations to text that was original to me, regardless of citation. All cited documents are in the public domain and should be available to the reader from a library or website, or by writing to the institution responsible for them.

I would like to thank the following for permission to quote passages: Cambridge University Press, the EU Delegation in Beijing, the IUCN SSC Conservation Breeding Specialist Group, *The Jakarta Post*, the Ministry for Foreign Affairs of Finland, the Swiss Agency for Development and Cooperation, the Swiss State Secretariat for Economic Affairs, *Asia and the Pacific Policy Studies*, and Virgin Books.

I dedicate this work to Mary, my partner in curiosity.

Chapter 1

Introduction

Complexity, curiosity, and fear

Those who craft aid investments seek to induce change within social, economic, and ecological systems. The investments themselves and their targets are very complex, and the interactions between them involve many individuals, institutions, and relationships, all influenced by attitudes, cultures, histories, and opportunities, whether consciously perceived by the actors or not. Despite complexity and uncertain outcomes, aid is a major human enterprise, which is driven by multiple objectives of which the headline aim is usually to reduce, and, eventually and sustainably to eradicate, extreme poverty. An amount equivalent to about 165 billion United States dollars (US$165 billion) each year is spent on overseas aid, and this is set to flow increasingly towards climate change adaptation as the consequences of our greenhouse gas (GHG) emissions become more harmful. Some of this money is also spent on reducing GHG emissions, but so far with little effect on the 50 billion tonnes or so emitted annually by our US$75 trillion world economy.

Over the last 15 years or so we have been trying to change our ways and ideas to allow for the diversity and resilience that we need to adapt to climate change, and the technologies and investments that we need to mitigate it by reducing emissions. This book is an eye-witness view of our experience in these 'baby-step' years, based on a synthesis of 50 aid evaluations across the developing world. These were done at all stages of the 'project cycle' (i.e. formulation, appraisal, inception, mid-term review, etc.), and assessed a total expenditure of over US$3 billion by five donors – the EC, Finland, Switzerland, Norway, and the UK – which together contribute more than a quarter of all aid. It concludes that we have learned enough to move to a higher level in designing investments and judging their performance. Implications for evaluation and training are spelled out, with examples from across the ecology, technology, and governance of climate change mitigation and adaptation, and linked areas of biodiversity and desertification.

The approach developed here is based on two premises. The first is that it is more often useful to respond to a complex system with curiosity than with fear. Curiosity is a wish to learn from what has happened and why, while fear is a wish to avoid being held accountable for mistakes. Curiosity makes us want to understand the whole system and how its parts respond to and influence one another, but it demands courage, imagination, and knowledge. Fear makes us want to break the system down into its parts, and measure them, yielding numbers and partial descriptions that are easily comprehended and therefore reassuring, but it demands simplification. This leads to the second premise, which is that when systems are very complex, as social, economic, and ecological systems are, a simplifying, reductionist approach can block understanding. This is to re-state a critique of what has been called the 'McNamara Fallacy' (Handy, 1995), which is named for its originator, Robert Strange McNamara, a systems analyst and corporate president at the Ford Motor Company who later held the roles of US Secretary of Defense (1961–1968) and President of the World Bank (1968–1981). His idea was to make planning easier by assuming that whatever can be measured easily should be measured, and that whatever cannot be measured easily should be disregarded. This line of thinking encourages a focus on metrics (most famously, the counting of dead Vietnamese people to quantify counter-insurgency efforts in the 1960–1975 Vietnam War) that distracts from efforts to understand complex systems. It is a seductive approach, because people are often reluctant to accept that highly complex systems are not amenable to easy, meaningful measurement. It therefore continues to shape investment in sectors such as medicine, education, and development. As well as encouraging the measurement of meaningless things, it can lead to the expectation that complex systems will behave as if they were simple, resulting in serious costs.

Taking these premises into account by encouraging curiosity and respect for complexity, the aim here is to improve aid performance and the climate change response. This is to be done through greater realism in describing, designing, and expecting change in the complex systems that aid investments seek to influence. Thus, in terms of evaluating aid performance, I intend to reverse the current direction of travel from curiosity towards fearfulness. At present, priority is often given to the few things that can be measured rather the many that cannot, and aid evaluators and planners tend to think in reductionist terms and train the next generation accordingly. Against this, I argue for building on an earlier approach developed by the Organisation for Economic Co-operation and Development (OECD) Development Assistance Committee (DAC), which is that multiple, complementary evaluation criteria should be used to examine aid investments. This calls for looking at complex systems from different angles and gathering evidence to shed light on each, so that each can shed light on the whole. The points of view concerned are the core criteria of relevance, efficiency, effectiveness, impact, sustainability, connectedness, coherence, and replicability, and other key considerations of partner satisfaction, cross-cutting themes, and design quality. By applying common definitions and scoring methods to these, at all stages of the project cycle, aid investments, individually and in the aggregate, can be understood, represented, and usefully compared with each other.

This approach can let the weakest investments in a portfolio be identified, corrected or cancelled, thus raising the portfolio's overall performance and value for money, and demonstrating this by way of its mean score. Such a score, however, would not be arbitrary, but defensibly rooted in evidence and a realistic description of all key dimensions of all investments. By these means, some major portfolios evaluated here have been shown to perform fairly well over the past 15 years, and more importantly to have improved their performance over time, due to better design and increasing attention to strategic factors such as climate change. Meanwhile, the potent effects of individual insight, sustained strategic effort, and historical, cultural, and accidental factors, have all been reliably demonstrated. The result is a body of applied knowledge, and a set of techniques to obtain, organise and present it, that can shift the trajectory of aid towards more valuable outcomes.

I further argue that the lessons to be learned from facing complexity should be treated as educational resources, and that the next generation should be prepared with subtlety and depth of experience to respond creatively to complexity. Some ideas are given in Chapter 9 for face-to-face courses to help convey the principles and practices of systemic aid evaluation and design. The need for these is especially urgent in the areas of ecological climate change mitigation and adaptation linked to preserving resilience and diversity. Overseas aid will be an important part of our collective efforts to mitigate and adapt to climate change, and the explanations and examples throughout this book should give the reader a clear sense of how to evaluate aid performance properly, and how to design aid investments for optimum performance under conditions of complex uncertainty.

Aims of the book

This book has two connected aims. The first is to explain how to evaluate the performance of aid investments, and by extension how to design them, so that aid money can be spent more consistently, wisely, and well, in building a richer and more equitable, rational, and stable world. This particularly applies in our efforts to mitigate and adapt to climate change, where the 2015 Paris Agreement foresees major aid investments and implies a need to improve their performance (see below). In line with this, the book makes a case for returning to a criteria-based approach to evaluation, building on the practices that were established in the 1990s. These have since become obscured by the inventiveness of those who commission evaluations, making it hard to compare investments and to identify patterns that can be used to enhance performance.

One particular challenge is that those who commission and practise aid evaluations have created an inconsistent vocabulary, and poorly explained principles and practices, that undermine clarity and transparency. These in turn limit efficiency within the profession, as well as its accountability to tax-payers, parliamentarians, and journalists. To help reverse this, I argue that the original evaluation criteria need to be increased in number, from five to eight, in order to address: (a) the phenomena of climate change, globalisation, mass extinction, and water crises; (b) the wasteful and contradictory practices of aid agencies; and (c) the lack of systematic learning from aid investments that stops successful ones being understood, scaled up, and replicated. Also needed is a consistent, evidence-based, and transparent scoring system to capture judgements on performance generated by evaluations, thus making it easier to detect important patterns of strength and weakness within and between aid portfolios.

The book's second aim is to respond to the fact that the world is itself changing fast. By 'world' I mean the biosphere, which comprises all the Earth's atmosphere, oceans and land surfaces, soils and underlying shallow rocks, and all of the living systems within them, including human populations, cultures and economies, and the soils, waters and landscapes that sustain them. Here there is much to concern us, including biodiversity loss and mass extinction, ecosystem and land degradation, drought and desertification, crises of water supply and sanitation, pollution of the land, sea, and air by a wide range of persistent and destructive toxins, and climate change. These problems are all large-scale, long-term, pervasive, resistant to piecemeal, easy, or cheap solutions, and in many cases mutually interactive and reinforcing. They also combine with others, such as overfishing, overpopulation, and war, to undermine the living conditions, livelihoods, health, and well-being of everyone on Earth.

Thus the long-term goals of human development are being undermined, with climate change in particular having the power to bring human development to a dead stop, and even to create a wholesale reversal of everyone's fortunes. It is clear that transfers of wealth in the form of overseas aid must comprise some part of our climate strategy, and the effectiveness with which this is spent is an important concern. This book is about the lessons learned from overseas aid, how to assess its effects, and how to improve its performance. The assumption is that these lessons and methods will be useful as aid is adapted to the emerging challenges of the next decades, a process in which high levels of performance and low levels of waste will be much in demand. Hence, as well as advocating reforms to the practices of aid evaluation, there is a need to offer a knowledge resource to help evaluators identify and understand the kinds of investment that can make most difference to sustainability. It is among these that evaluators will be looking when seeking evidence upon which to base judgements on the performance of aid investments.

Combining the two aims, the book is for making aid investments better in their own terms, *and also* for making them better at contributing to ecological sustainability during this global crisis. Hence the book is intended to be useful across the entire aid profession, from students and trainees, to ministers and officials in government departments, and to the consultants employed directly or indirectly by those departments, who help deliver the policies, projects, programmes, partnerships, designs, and evaluations that together facilitate all aid expenditure. The underlying assumption is that many of these people would like to learn what works, what does not, and why, partly in order to justify what they have done but also, and far more importantly, to systematically improve the performance of the aid activities for which they are responsible.

Policies guiding overseas aid

Polices are the prior decisions that participants in an aid investment have already made, with the implication that they will shape priorities and decisions on what to do, how to design it, and how to assess its worth, so they have practical implications for aid evaluation and design. Thus, for example, the alignment of an investment with relevant policy is a key reference point in assessing its relevance. Brief summaries of the main points of development policies put out in the late 2000s by several donor countries are given in Box 1.1, all being considered 'like-minded', and all presumably being influenced by, and their predecessors contributing to, the European Consensus on Development (EU, 2005). These development policies, taken together, amount to a unique and inspiring common expression of values and purpose.

Box 1.1 Policies of some European 'like-minded' donors in the late 2000s

Denmark. Denmark's development policy was adopted by Parliament in June 2010 and is an integral part of the country's foreign policy. Like its previous policies, it claims to be rooted in universal human rights and respect for fundamental freedoms, as they are set out in the UN Charter and the Universal Declaration of Human Rights. On page 4 it observes that "Fighting poverty is about giving each individual the opportunity and the ability to influence their own situation in life, politically, economically and socially." Its five priority areas are: (a) growth and employment, (b) freedom, democracy and human rights, (c) gender equality, (d) stability and fragility, and (e) environment and climate. It responds to the need to step up efforts to reach the Millennium Development Goals (MDGs), to proactively address emergent challenges (such as climate change, food security, water security, ecosystem degradation and biodiversity loss), and strongly emphasises the importance of partnerships in the fight against poverty. It reiterates a primary focus on Africa and on improved policy coherence, and calls for stronger engagement in fewer countries, and a stronger role for the EU in development cooperation. It asserts the aim of making the best use of Danish competencies to produce results that are well documented and communicated.

Sweden. With the overriding goal of combating poverty and promoting equitable and sustainable development, Sweden's development policy is built around the perspectives of poverty reduction and respect for human rights. It makes a commitment to effective, high-quality development cooperation, emphasises coherence across all policy areas, and embraces a global approach. A recent focus on Africa identified three thematic priorities for development cooperation: (a) democracy and human rights; (b) environment and climate change; and (c) gender equality and the role of women in development. A recent reform process included reducing the number of partner countries, streamlining thematic policies and guidelines, and strengthening evaluation processes, and aimed to improve the efficiency and focus of Swedish aid, and the quality of knowledge and communication about its poverty reduction impacts.

Norway. The Government's 2009 white paper on development policy outlines Norwegian development policy objectives in a rapidly changing world, where climate change and conflicts are making the fight against poverty harder. It reconfirms a commitment to the MDGs, to a rights-based development policy and to a UN-led world order and identifies climate change, violent conflict and lack of capital to be the most important challenges in the fight against poverty. Norway is committed to continue its efforts in key areas where it feels it can provide added value, such as sustainable development, peace building, human rights and humanitarian assistance, oil and clean energy, women and gender equality, good governance and the fight against corruption. The policy raises three elements as crucial for development – a functioning state, an active civil society, and a viable private sector – and highlights a set of approaches and policy measures that should provide a coherent and effective framework for Norwegian development cooperation. These include encouraging partnerships with the private sector with a view to increasing investment in poor countries; working to promote an international system outside the traditional framework for official aid to fund global public goods; cooperating more closely with immigrant communities in Norway on remittance transfers to developing countries; and concentrating assistance in areas where Norway has special expertise that is in demand. These include peace building, managing revenues from natural resources and promoting women's empowerment, as well as an overall aim to shift the focus of development assistance increasingly towards countries emerging from violent conflict and countries facing particular problems due to climate change. Through its climate and forest initiative, Norway has taken a leading role internationally in climate change mitigation and protecting tropical forests, which can be seen as part of Norwegian overall efforts to secure global public goods.

The Netherlands. Dutch development policy gives priority to: (a) economic growth and equity; (b) security and development; (c) equal rights and opportunities for women and girls; and (d) climate, sustainability and energy. This policy is driven by the concern that the international community is lagging behind in several areas of the MDGs, and 'Project 2015' directs Dutch priorities to the regions that are lagging furthest behind globally in achieving the MDGs. The policy stresses policy coherence, noting that the MDGs are a concern not only for development cooperation but for everyone (i.e. for all ministers, the private sector, and society as a whole). Within this strategy, Dutch partner countries have been grouped into three categories, one of which specifically focuses on speeding up achievement of the MDGs. Other policy documents stress the need to intensify the fight against corruption, to open up the aid industry through greater complementarity between existing players, and to improve the effectiveness of aid. There is active reporting and communication of development challenges and results achieved.

The United Kingdom. The International Development Act 2002 stipulates poverty reduction, in particular through achievement by 2015 of the MDGs, as the overriding goal of development assistance. Two government White Papers (in 2006 and 2009) built upon this policy framework and adopted a comprehensive approach which goes beyond the aid agenda to address new global challenges. The 2009 White Paper defines four key priorities for UK aid: (a) achieving sustainable growth in the poorest countries; (b) combating climate change; (c) supporting conflict prevention and fragile states; and (d) reinforcing the international aid system's efficiency and effectiveness. The UK is particularly concerned with humanitarian assistance and reform of the international aid system, and expresses a strong commitment to policy coherence, to ensure that all of its domestic and international policies support, or at least do not undermine, partner countries' development aspirations. The 2009 White Paper provides for coherence around three key priorities: poverty reduction and economic growth (including trade), climate change, and conflict. The International Development (Reporting and Transparency) Act 2006 obliges the UK's Department for International Development (DFID) to report annually on the impacts of UK policies on development, and an OECD/DAC Peer Review suggests that there is room for improvement in the measures, monitoring and reporting (to parliament and the public) on the poverty reduction impact of Britain's domestic and foreign policies.

The European Union. The EU acts both as a grouping of Member States and as a unified body. In practice this means that the EU acts as one entity on some issues (such as trade), but works in parallel with the Member States on others (such as aid). The European Consensus on Development officially defines common objectives and principles that are shared by all Member States, emphasises poverty reduction as the central goal, and expresses a commitment to aid effectiveness and policy coherence. It identifies a number of priority areas, including water and energy, food security, conflict prevention and fragile states, rural development and regional integration, with cross-cutting themes including promotion of democracy, gender and environmental sustainability. There is a momentum for change in the context of the ratification of the Lisbon Treaty (which highlights the importance of sustainable development and poverty reduction in the EU's external relations), with a stronger parliament and the creation the European External Action Service in 2010. This might be expected to create greater coherence and synergy among the Commission and the Member States across the areas of trade, foreign, security and development policy.

Sources: Appendix 1 (10–31); reviews of UK (2002), MFA Sweden (2003, 2008), EU (2005), MFA Netherlands (2007, 2009), MFA Norway (2009), ETTG (2010), Gavas *et al.* (2010), Norad (2010), and MFA Denmark (2010) by Caldecott *et al.* (2010: 88–91).

Aid policies evolve, partly through occasional agreements such as the December 2015 Paris Agreement that was made under the United Nations Framework Convention on Climate Change (UNFCCC). This seeks to guide the climate change response at an international level, in terms of mitigation, adaptation, and finance, from the year 2020. The UNFCCC (2016) description of the Paris Agreement is summarised in Box 1.2, but it should be noted that other observers stress slightly different things – for example, the EC Climate Action (2016) summary emphasises that agreed plans for national GHG emission reductions are inadequate to meet goals on temperature rise, and notes the commitment of developed countries to mobilise US$100 billion per year for the climate change response by 2020, while a higher goal is to be set for later. In any case, the Paris Agreement is clearly an important step forward, and provides a major part of the global context for many of the aid efforts to which this book seeks to contribute.

Box 1.2 The 2015 Paris Agreement on climate change action

The Paris Agreement was made at the 21st Conference of the Parties to the UNFCCC in December 2015, and entered into force in November 2016. Its essential elements include:

- *Long-term temperature goal* – the increase in global mean temperature above pre-industrial levels will be below 2°C during the 21st century.
- *Global peaking* – GHG emissions will peak as soon as possible for each country, with the aim of achieving a global balance between emissions and removals in the second half of the century.
- *Mitigation* – each Party will prepare a 'nationally determined contribution' to the worldwide goal and try to put it into effect, reporting on progress every five years.

- *Sinks and reservoirs* – all Parties are encouraged to conserve sinks and reservoirs of GHGs, including forests.
- *Adaptation* – national adaptation efforts (i.e. to increase adaptive capacity and resilience, and reduce vulnerability) will be strengthened through support and international cooperation.
- *Loss and damage* – ways will be sought to help vulnerable countries cope with the adverse effects of climate change, including extreme weather events and sea-level rise.
- *Support* – developed countries will support the efforts of developing countries to build clean, climate-resilient futures, including through cooperation on climate-safe technology development and transfer, and building capacity, with the aim of a balance between adaptation and mitigation.
- *Transparency* – robust transparency and accounting systems will be used, and information submitted by each Party will undergo international review.
- *Global stocktake* – a 'global stocktake' will take place in 2023 and every five years thereafter, to assess collective progress and enhance action and cooperation.

The agreement encourages action before 2020, especially on technical cooperation, meeting urgent needs, and high-level engagement. It also validates climate change efforts by civil society, the private sector, financial institutions, cities and other sub-national authorities, recognises the need to strengthen the knowledge, technologies, practices, and efforts of local communities and indigenous peoples, and notes the important role of incentives provided through tools such as domestic policies and carbon pricing.

Source: UNFCCC (2016).

Dimensions of overseas aid

Overseas aid is a significant activity for the governments of donor and recipient countries, and also for the many people and institutions involved in managing and spending it, for profit or not. Official Development Assistance (ODA) comprises government spending on deliberate efforts to promote economic development and well-being, and also financing offered by governments to other governments on concessional terms for these purposes. With all figures converted into constant 2011 United States dollar equivalents, the countries that belong to OECD/DAC allocated about US$135 billion to ODA in 2013 (Table 1.1). At the time, their total gross national income (GNI), which is the same as their gross domestic product (GDP) but corrected for net debt transactions, was about

Table 1.1 GDP/GNI and ODA by OECD/DAC member country

OECD/DAC member country	*Total gross domestic product (US$ trillions, 2013)*	*Official development assistance (ODA) (US$ billions, 2013)*	*ODA as share of gross national income (%, 2013)*
Australia	1.56	4.85	0.34
Austria	0.43	1.17	0.28
Belgium	0.53	2.28	0.45
Canada	1.83	4.91	0.27
Czech Republic	0.22	0.21	0.11
Denmark	0.34	2.93	0.85
Finland	0.27	1.44	0.55
France	2.81	11.38	0.41
Germany	3.73	14.06	0.38
Greece	0.24	0.31	0.13
Iceland	0.01	0.04	0.26
Ireland	0.23	0.82	0.45
Italy	2.15	3.25	0.16
Japan	4.92	11.79	0.23
Korea (South)	1.31	1.74	0.13
Luxembourg	0.06	0.43	1.00
The Netherlands	0.85	5.44	0.67
New Zealand	0.19	0.46	0.26
Norway	0.51	5.58	1.07
Poland	0.53	0.47	0.10
Portugal	0.23	0.48	0.23
Slovak Republic	0.10	0.09	0.09
Slovenia	0.05	0.06	0.13
Spain	1.39	2.20	0.16
Sweden	0.58	5.83	1.02
Switzerland	0.69	3.20	0.47
United Kingdom	2.68	17.88	0.72
United States	16.77	31.55	0.19
Total (mean)	*45.21*	*134.85*	*(0.30)*

Sources: OECD (2014); World Bank (2015).

US$45 trillion, so their average ODA allocation was equivalent to about 0.3 per cent of total GNI.

Other sources of ODA included the EU Institutions, which gave a further US$16 billion or so, and non-DAC members (United Arab Emirates, Turkey, Russia, Israel, Hungary, Estonia and Latvia), which gave a total of almost US$10 billion between them, or 0.25 per cent of their GNI overall. Meanwhile, Saudi Arabia, India, Venezuela, China, and others, have also contributed ODA in recent years, and these and many other countries provide investment guarantees and other subsidies and grants (some of them invisibly) in line with their trade, ethnic, religious, security, and other concerns. And there are also several billions of dollars in charitable donations annually, organised by international institutions that focus on humanitarian, compassionate, environmental, and development activities. Thus, in the context of a US$75 trillion world economy, the equivalent of around US$165 billion each year is conveyed outside normal market channels for purposes of 'development'. All of this is loosely called 'aid' here, although variants of the phrase 'development cooperation financing' are widely preferred, and those described as 'donors', referring to the sources (whether proximate or ultimate) of aid, often prefer to call themselves 'cooperation partners' or some such. In any case, whether the total amount of aid is 'a lot' or 'a little', and whether it is being spent on the 'right' things, depends on who is judging, and there are many points of view.

Main knowledge resources for the book

Large statistical databases on aid projects do already exist. For example, the European Commission (EC) has developed a results-orientated monitoring (ROM) system: "ROM reviews are conducted by independent experts through regular onsite assessments of projects and programmes in all EU partner countries. A consistent, highly structured methodology ensures the quality and the comparability of the collected data. Projects and programmes are given simple scores against internationally agreed criteria (relevance, efficiency, effectiveness, impact, and sustainability) substantiated by concise explanatory texts" (EuropeAid, 2012: 8). The ROM system is complemented by recommendations on how to improve performance. More than 2,000 ROM reports were prepared in 2010 alone, with the data being captured by the Common RELEX Information System (CRIS), in principle to provide "an overview of the performance of the EU development aid portfolio" (*ibid.*).

Such an overview based on pre-digested indicators would support certain managerial and reporting tasks at the macro level, but in this book my aim is to unpack a sample of investments for investigation at a deeper level, starting with the projects, programmes and partnerships themselves, treated as complex systems. Here the added value will come from direct familiarity with specific cases. Thus the book is based on a sample of 50 evaluations in which the author was involved, covering nearly two dozen aid modalities in 150 or so countries (Appendix 1). Most of these investments were by the EC, Finland, Switzerland, Norway, and the UK, which together contributed about 27 per cent of all aid in 2013 (Table 1.1). All these donors have reputations for being innovative in their thinking and in the approaches that they are willing to try, but it is possible that strategies not mentioned here exist within the portfolios financed by other countries. Although the analysis here is largely based on bilateral (i.e. country-to-country) donations, even these often involve multilateral (i.e. multi-governmental) channels, and the principles that emerge are just as relevant to international financial institutions such as the World Bank as they are to solo donors, groups of like-minded donors, and non-governmental (charitable) donors.

These principles may be stated simply enough:

- that some methods, arrangements, or modalities perform better than others, because they *promote motivation and capacity among practitioners*, and some perform worse because they do not;
- that some rationales or approaches perform better than others, because they are *based on sound situational analyses and wise theories of change*, and some perform worse because they are not; and
- that political contexts can change drastically, so aid *agencies must be light on their feet* when they need to be, rather than playing catch-up years later.

This book uses examples, both of arrangements and of approaches, to show how institutions have organised their people and tasks in the recent past, how their performance can be assessed systemically, and how the actors involved can reliably take the better rather than the worse path in an increasing range of circumstances. The aim is to make it possible for performance to improve in real terms, particularly in those fields where poverty, water, deserts, biodiversity, and climate change intersect in various combinations.

Development of the evaluation approach taken here

The 2002–2009 project evaluations

In evaluations led by the author in 2002–2009, for EU Delegations in Bangkok, Beijing, Jakarta, and Manila (Appendix 1: 2–4, 6, 9), the five original DAC criteria of relevance, effectiveness, efficiency, impact and sustainability were taken as given, but the terms of reference (ToR) also instructed the consultant to "verify, analyse, and assess the integration and impact of cross cutting issues" and to use "professional judgement and experience to review all relevant factors" (e.g. EC, 2007: 4). These instructions were interpreted to request a more systemic evaluation than might otherwise have been done, but the lack of specificity in the ToR opened the way for other evaluations by other teams to vary in their interpretation, at the cost of comparability among them. In any case, these specific evaluations did not stray far from reporting on the five DAC criteria, but in the process the need for additional criteria became apparent, and observations on connectedness, coherence and replicability were explored in reports under a variety of headings.

The 2010 synthesis evaluation

The ToR for the synthesis evaluation of the 'Sustainability Dimension in Addressing Poverty Reduction' (Appendix 1: 10–31), which was commissioned by the Finnish Ministry for Foreign Affairs (MFA Finland), similarly stated that "The evaluation will utilize the five OECD/DAC development evaluation criteria, relevance, efficiency, effectiveness, sustainability, and impact, as well as the additional criteria of coherence and consistency, complementarity, and coordination, compatibility, and the Finnish value-added, as appropriate... Due consideration must be given to the different tiers of development, the global policy goals (including, MDGs, the Paris Declaration and the Accra Platform) and specific concerns (including Climate Change and Adaptation and related Disaster Risk Reduction), development policies at donor and at partner country levels, including cross-cutting themes, as well as to the implementation of policies through practical development cooperation interventions" (MFA Finland, 2009: 6).

This freight of additional evaluation purposes drew attention to the need to define additional criteria, since a way had to be found to make sense of hundreds of projects within a portfolio organised through numerous modalities and aid instruments. I therefore devised a suite of 14 indicators to capture as many as possible of all the dimensions of interest (Table 1.2), and led a review of all 22 thematic and geographic programmes from all 14 points of view, summarising the key findings with respect to each, and scoring them consistently by performance according to the evidence found in each case. The results were encouraging, in that defensible, evidence-based patterns in performance were demonstrated that could be explained in terms of the behaviour of institutions and individuals in complex and dynamic circumstances. It was found that investments consistently performed better according to some criteria than to others, some reasons for this could be understood and explained, and specific recommendations could be made to improve those parts of the system that consistently produced weak outcomes.

The 2011–2012 country studies

The opportunity arose to apply the lessons of the 2010 synthesis evaluation to decade-long Finnish cooperation programmes with Nepal, Nicaragua, and Tanzania (Appendix 1: 33–35). Again commissioned by MFA Finland, the ToR likewise stated that "The evaluation will use the development evaluation criteria, relevance, effectiveness, efficiency, sustainability and results/effects in the longer-term, and the additional criteria of coordination, cooperation, coherence and Finnish value added. The criterium [*sic*] 'impact' has deliberately been left outside, because the purpose of this evaluation is not to scrupulously examine each individual intervention, but rather the entire programme that these interventions constitute" (MFA Finland, 2011: 5). To reflect the strategic and partnership-based nature of the country programmes, the *effectiveness* criterion was replaced by that of *strategic effectiveness*, and *activity design* by *programming logic*, but the 14 criteria were otherwise again used to examine the programmes in the context of the whole country as a developing system influenced by numerous cultural, historical, ecological, economic, and political factors, to which the Finnish programmes had had to adapt over time. Again the results

Table 1.2 Definitions of the 14 criteria used in the 2010 synthesis evaluation

Criterion	*Description*
Relevance	Relates to problems and needs, and concerns whether the results, purpose and overall objectives of the activity are in line with the needs and aspirations of the beneficiaries, and with the policy environment of the activity.
Efficiency	Relates primarily to sound management and value for money.
Effectiveness	Relates to the achievement of results that further progress towards achieving the activity's specific purpose.
Impact	Relates to wider effects, and therefore the extent to which durable changes in any aspect of the well-being of targeted beneficiaries or others have improved as a direct or indirect result of the activity.
Sustainability	Relates to the likely continuation of results and impacts achieved by the activity after it ends.
Coordination	Also known as harmonisation, coordination describes the interaction with relevant groups and other donors in a partner country, ideally so that synergies occur and conflicts or overlaps do not.
Complementarity	Relates to how well concurrent activities support one another, and the specific skills and benefits that various partners in an activity or a region can bring to achieving desired outcomes.
Compatibility	Also known as alignment, compatibility relates to how well the goals of Finland's development cooperation policy or partner country's development policy are taken into account in planning and implementing activities.
Connectedness	Relates to the linkages between systems that are being targeted by an activity and other systems that may affect outcomes (i.e. vulnerability or resilience to external factors).
Coherence	Describes whether activities are in line with internal policies and strategies, and in harmony with those of other ministries involved in development cooperation.
Finnish added value	Describes the contribution to an activity of knowledge, skills, approaches, priorities and processes that are specifically Finnish in nature.
Partner satisfaction	Refers to the extent to which the activity's partners are satisfied with its results, equally applied whether those partners are global, regional, national, local or community-level institutions.
Activity design	Relates to the analysis of context, problems, needs and risks, and the assembly of measures that can realistically, adaptively and efficiently drive the activity to deliver useful results and sustainable impacts.
Replicability	Concerns the learning of lessons, the understanding of processes and relationships, and the possibility that future activities can be modelled on improved versions of past ones.

Sources: Appendix 1 (10–31); Caldecott *et al.* (2010: 41–42).

were encouraging in terms of method, but the greater amount of information available at country-programme level drew attention to its patchy distribution across the evaluation criteria and the fact that too few data were available to support subtle distinctions among certain criteria (e.g. *coherence*, *compatibility*, *coordination*, and *complementarity*).

The 2013–2014 climate effectiveness study

A study of the large portfolio of projects financed in 2000–2012 and judged to be relevant to climate change by the Swiss government offered scope to refine evaluation methods in the areas of plausibility, evidence, and scoring (Appendix 1: 37–46). This evaluation focused on effectiveness, the highest-profile criterion for a naïve audience, and the ToR defined its main purpose to be "informing the interested public and the Members of the Swiss Parliament about the effectiveness of the Swiss International Cooperation" (SDC & SECO, 2013: 3–4). With 508 projects and a total spend of over US$1.4 billion to assess, the first task was to summarise the approach of each project and, using the Rio Climate Markers (see Chapter 3), to compare it with accepted international criteria of what constitutes a climate-relevant project. The aims here were to validate the project's inclusion in the portfolio (i.e. to check that Swiss investment in meeting the climate change challenge had not been exaggerated), and to classify it by theme according to the relevant sub-set of possible mitigation and adaptation approaches. Enough evidence was then obtained from all sources to form judgements on the likely effectiveness of 423 of the projects covering 92 per cent of the total budget, and 61 were examined in detail through field work, desk studies, and interviews as a way to 'ground truth' the findings and confirm the scoring results. In the process, the criterion of *activity design/programming logic* was replaced by a separate analysis of *project design*, in which the project's relevance to climate change and the general quality of its design were assessed and scored (Table 1.3). Detailed findings are used in this book, drawn from the technical report of the evaluation (Caldecott *et al.*, 2014), including its annexes on: methodological guidance (*ibid.*: 73–79); field analyses of 30 projects (*ibid.*: 110–269); desk reviews of selected projects in Vietnam (*ibid.*: 270–307) and 25 other projects (*ibid.*: 308–401); and the effectiveness of the World Bank, Inter-American Development Bank, and UNDP (*ibid.*: 406–411). A version of this report for public and media use is also available in SDC & SECO (2014), and there is a summary and discussion in Kissling & Windisch (2017).

Other evaluations

In evaluations during 2015, such as those of the Finnish programme in Kenya (Appendix 1: 48) and the Forest Carbon Partnership Facility (Appendix 1: 50), redundancies and ambiguities were removed from the definitions of evaluation criteria, further ideas on research practices (Robson, 1993; Mayne, 2011; Ritchie

Table 1.3 Guidelines for presenting evidence on climate change effectiveness

Section	*Guideline*
Identification	Number (SDC 7F-00000 or SECO UR/UZ 00000) and title of project.
Documents used	Use the credit proposal for the 'plausibility of project design' section. Use the final reports and external summative evaluations (if available) for the 'direct evidence of effectiveness' section. Refer to any other documents that may have been used, for example in assembling indirect evidence for effectiveness.
People interviewed	Name and institutional position of each person interviewed (their observations to be integrated into the 'evidence' and/or 'overall effectiveness' sections).
Basic data	Start/end date, budget/disbursements.
Location	Be as specific as possible, and include a brief description of the socioeconomic and ecological context.
Partners	Identify and briefly explain the roles of all institutions involved in the project.
Result chain (RC)	Define the RC to which the project has been assigned by SDC/SECO, the nature of the pathway, the output(s) and outcome(s) involved, and the validation criteria that are expected to apply to it.
Purpose	A very concise yet comprehensive, accurate and defensible summary of what the project is or was for, based on the original credit proposal.
Pre-review estimates of relevance to climate change (CC)	Summarise how the project was assessed by SDC/SECO (per cent relevant to adaptation/mitigation), and how it was initially classified by the review team, including the specification of any validation criteria that it was judged to meet and the basis for such a decision.
Evidence for climate change mitigation and/or adaptation effectiveness	
1. Evidence for direct CC effectiveness of the project (GHGs reduced, adaptation)	This section is for summarising evidence of CC mitigation and/or adaptation effectiveness from the project itself. This evidence can be direct and quantitative or qualitative, but should not be speculative.
2. Evidence of indirect effectiveness of the project (side effects, other consequences)	This section is for summarising other information that is relevant to forming a judgement on the likely indirect climate change effectiveness, any CC-relevant side effects, and expected/unexpected consequences of the project.
3. Reasons to expect CC effectiveness of this kind of project based on other knowledge	This section is for summarising any potential further evidence based on similar kinds of projects in the same country or other parts of the world to build a case for or against the likely effectiveness of this particular project.

Table 1.3 continued

Section	*Guideline*
Overall conclusion on effectiveness based on the evidence (1 + 2 + 3)	With explicit reference to the evidence, whether direct (quantitative or qualitative) or indirect, and specifying what kind of effectiveness is involved, score the project as a whole as: 'a' (extremely strong = 7), 'a/b' (very strong = 6), 'b' (strong = 5), 'b/c' (moderate = 4), 'c' (weak = 3), 'c/d' (very weak = 2) or 'd' (none = 1). Note that the overall effectiveness score is a judgement based on the evidence about effectiveness, and must be defensible using that evidence or reasonable inferences from it.
Project design aspects	
CC relevance of project design (evidence and reasoning, integrity of the RC pathway)	This provides an opportunity to consider and critically evaluate the processes by which it was decided to invest Swiss public money in the project. If there is no evidence the reviewer should say so. If there is enough evidence to form a judgement on any aspect of the plausibility of project design, that aspect may be scored as: 'a' (excellent = 7), 'a/b' (very good = 6), 'b' (good = 5), 'b/c' (adequate = 4), 'c' (problematic = 3), 'c/d' (poor = 2) or 'd' (seriously deficient = 1). For *evidence and reasoning*, describe and score the empirical evidence and reasoning upon which the logical pathway from CC challenge to response is based, as articulated in the credit proposal (a high score is given where plausible reasoning builds on sound evidence). For *pathway integrity*, describe and score the closeness with which the steps of the logical pathway from CC challenge to response are followed in the credit proposal (a high score is given where there are strong connections between steps).
General quality of project design (clarity of explanation, extent of participation)	For *explanation clarity*, describe and score the clarity with which the logical pathway from CC challenge to response and the choices within it are explained in the credit proposal. For *participatory design*, describe and score the extent to which local research and consultation processes contributed to the design of the project. It is assumed that building on local knowledge and participation will yield a more reliably effective project than not doing so, but if there is any evidence to the contrary (e.g. that the project is so obviously 'no regrets' that little local input was needed), the reviewer should say so and not score this aspect (it will then need to be discussed in the report).

Sources: Appendix 1 (37–46); Caldecott *et al.* (2014, Annex 2: 73–75).

et al., 2014) were considered, and a refined set of eight criteria were developed for use in future evaluations (Table 1.4). As with the Swiss portfolio evaluation, the concept of design quality, based on the clear articulation of logic and the process by which investments were designed, was retained in an analysis separate from the evidence-based approach applied to the criteria themselves.

Table 1.4 Guidelines for judging performance and design quality

Criterion	*Scoring performance*
Relevance	A high score is given if a design responds to the needs of the beneficiaries in their political, economic and ecological contexts, and where it is aligned with the overall policy environment through a convincing theory of change.
Efficiency	A high score is given if a design contains measures that through elegance and accountability promote sound management and value for money, yielding confidence through counterfactual ('what if not') analysis that the same or better results are unlikely to be achieved through different means or with lower overall expenditure or with different rates of expenditure.
Effectiveness	A high score is given if a design is specific in explaining how results to be obtained will contribute to achieving the project's purpose, usually specified in terms of the project itself but which may include contributions to broader cross-cutting objectives and other policy goals.
Impact	A high score is given if a design is specific in explaining how and why the project is likely to have effects that are wider and longer-term than its results.
Sustainability	A high score is given if a design is specific in explaining how and why the project is likely to have effects that will continue because of induced irreversible change in laws, institutions, relationships, public opinion, etc.
Connectedness	A high score is given if a design is specific in explaining how and why the project is vulnerable to external factors and influences over which it has little or no control, and how these influences are to be mitigated.
Coherence	A high score is given if a design is specific in describing how the project will maximise opportunities for synergy with, and manage interference from, the plans and actions of other actors. Factors include: *compatibility* (i.e. how well the goals of all participants are taken into account and where necessary reconciled); *coordination* (i.e. the existence and likely use of forums to sustain dialogue among stakeholders); and *complementarity* (i.e. how well participants' policies, plans, actions and choices support one another, and the degree of harmony among partners in achieving desired outcomes).
Replicability	A high score is given if a design is explicit in explaining the potential for the project to yield lessons that can be used to improve actions in the future or elsewhere.
	Scoring design quality
Evidence and reasoning	A high score is given if a design provides a convincing analysis of the context, problems, needs and risks upon which it is founded, and presents sufficient evidence that its approach can deliver useful results and sustainable impacts.
Clarity of explanation	A high score is given if a design conveys a complete sense of how and why the project should be implemented.
Participation in design	A high score is given if it is clear that all key stakeholder groups have been consulted in the design process, and have influenced it enough that their support can plausibly be expected.

Sources: Appendix 1 (50); Caldecott *et al.* (2015: 39–40).

The use and implications of the core evaluation criteria are explored in Chapters 4–6, but other influences on aid performance were also studied, including how aid portfolios might be assessed, analysed, disaggregated, and their elements compared. Findings are provided throughout this book, all drawing on the sources listed in Appendix 1. They allowed further exploration of: the influence of aid policies; the role and qualities of partnerships; cross-cutting themes and mainstreaming issues; the use and utility of design quality criteria; institutional and other factors in evaluations and ways to accommodate them; how to anticipate and compare performance over time and among investments; how to use aggregate performance statements; and how to induce transformations and recognise opportunities arising from them (see Chapters 2, 7, and 8).

An evaluation of the World Bank Group's GHG-reduction portfolio (World Bank, 2010) gave some additional perspectives. It concluded that the Bank had multiplied ten-fold in six years its annual investments in renewable energy and energy efficiency (to US$2 billion) and had helped mobilise more than US$5 billion in concessional funds for GHG reduction. But it also made the point that the Bank's "resources are small compared with the multitrillion dollar investments needed for low-carbon growth", which it took as a reason to ask "*How can the Bank have the greatest impact, both for development and for GHG mitigation?*" with its available resources. This sense of the challenge being far greater than available resources pervades the climate change issue, along with the thought that public funds must offer catalytic ideas, experiments, seed funding, and guarantees, to allow economy-wide transformations driven by private investment. But solutions cannot only be about money, since: first, private investment requires close supervision if it is not to create harmful impacts; second, social mobilisation on the scale needed requires leadership and grass-roots enthusiasm which cannot be bought; and, in any case, third, the scale of collective behavioural change required for us to make peace with nature (Box 1.3) will depend on a degree of cultural evolution. There are substantial tensions between these processes and motivations.

Box 1.3 Declaring 'peace with nature'

"In 1948, Costa Rica abolished its armed forces and turned its barracks into classrooms for its children. Almost sixty years later, in Costa Rica and around the world, another type of 'war' is raging, this time with nature, and the Government of Costa Rica decided that another declaration of peace was needed: a declaration of Peace with Nature, and the abolition of the forces that destroy it. By 2007, Costa Rica had spent nearly 20 years in an attempt to achieve environmental as well as social sustainability, using a variety of instrument mixtures that included rationalisation and expansion of the protected area system, debt-for-nature swaps to reduce public indebtedness, and leveraged investment into biodiversity conservation, reform of forest laws to encourage reforestation, institutional reform to enhance performance against new goals and targets, promotion of nature-oriented tourism, and organising new revenue-earning arrangements from the use of the nation's biodiversity." In 2007, President Óscar Arias Sánchez announced four national Peace with Nature commitments by the Government of Costa Rica:

- "that Costa Rica will become entirely carbon neutral by the year 2021, with all emissions offset by forest protection or reforestation at rates calculated by the National Fund for Forestry Financing (FONAFIFO), complemented by other actions including a significant reduction in the tax on hybrid automobiles and other vehicles that use renewable energy;
- "that with the immediate signing of an Executive Order requiring all State institutions to design and implement an environmental action plan, a clear signal will be given that Peace with Nature 'is not just tangible, but obligatory; not just of the people, but led by the State';
- "that there will be an increase in forest cover and the size of protected areas, with the expansion of the system of Payment for Environmental Services through FONAFIFO to cover an additional 6,000 sq. km of forest, the planting of five million trees in 2007, and the expansion of systems of biological corridors to promote gene flow among conservation areas and between them and surrounding landscapes; and
- "that a curriculum of sustainable development and environmental education will be actively promoted in all schools, 'to make sure our children and young people grow accustomed not to subtracting trees, but adding forests, multiplying mangroves and giving all the species within our borders an exponential value'."

Source: Caldecott *et al.* (2008: 118).

References

Caldecott, J., Chambers, B. & Kurukulasurya, L. (2008) *Green Breakthroughs: Solving Environmental Problems Through Innovative Policies and Law.* United Nations Environment Programme (Nairobi).

Caldecott, J. O., Halonen, M., Sørensen, S. E., Dugersuren, S., Tommila, P. & Pathan, A. (2010) *Evaluation of Sustainability in Poverty Reduction: Synthesis.* Evaluation Report 2010: 4. Ministry for Foreign Affairs (Helsinki).

Caldecott, J. O., Dornau, R., Gollan, J., Halonen, M., Saalismaa, N., Schilli, A., Simonett, O., Stuhlberger, C. & Tommila, P. (2014) *Technical Report on Effectiveness of the Swiss International Cooperation in Climate Change Mitigation and Adaptation Interventions 2000-2012.* Mandate of the Swiss Agency for Development and Cooperation (SDC) and the State Secretariat for Economic Affairs (SECO) (Bern).

Caldecott, J., Robledo, C. & Clarke, M. (2015) *The World Bank: Second Program Evaluation of the Forest Carbon Partnership Facility (FCPF), Inception Report.* Indufor Oy (Helsinki) and Forest Carbon Partnership Facility (Washington, DC). www.forestcarbonpartnership.org/sites/fcp/files/2015/November/FCPF%20Inception%20report%20Final%20Nov%202015.pdf (consulted 26 Mar 2016).

EC (2007) *Terms of Reference: Final Evaluation of Coastal Habitats and Resources Management Project (CHARM), Thailand (THA/RELEX/2000/0050).* EU Delegation (Bangkok) and European Commission (Brussels).

EC Climate Action (2016) *Paris Agreement.* ec.europa.eu/clima/policies/international/negotiations/paris/index_en.htm (consulted 12 Sep 2016).

ETTG (2010) *New Challenges, New Beginnings – Next Steps in European Development Cooperation.* European Think Tanks Group (including Overseas Development Institute, London; Deutsches Institut für Entwicklungspolitik, Bonn; Fundación para las Relaciones Internacionales y el Dialogo Exterior, Madrid and European Centre for Development Policy Management, Maastricht). Overseas Development Institute (London).

EU (2005) *Joint Statement by the Council and the Representatives of the Governments of the Member States Meeting within the Council, the European Parliament and the Commission on European Union Development Policy: 'The European Consensus'.* General Secretariat Note 14820/05 (DEVGEN 229, RELEX 678, ACP 155). Council of the European Union (Brussels).

EuropeAid (2012) *ROM Handbook: Results-Oriented Monitoring.* Support to Quality Monitoring Systems and Methodologies of Projects and Programmes of External Assistance Financed by the European Community (EuropeAid/127731/C/SER/Multi EVA/219719). European Commission (Brussels).

Gavas, M., Maxwell, S. & Johnson, D. (2010) *Consolidation or Cooperation: the Future of EU Development Cooperation.* Deutsches Institut für Entwicklungspolitik (Bonn).

Handy, C. (1995) *The Empty Raincoat: Making Sense of the Future.* Arrow (London).

Kissling, M. E. & Windisch, R. (2017) Lessons from Taking Stock of 12 Years of Swiss International Cooperation on Climate Change. In *Evaluating Climate Change Action for Sustainable Development* (edited by J. I. Uitto, J. Puri & R. D. van den Berg). Springer (Berlin).

Mayne, J. (2011). Addressing Cause and Effect in Simple and Complex Settings through Contribution Analysis. In *Evaluating the Complex* (edited by R. Schwartz, K. Forss, and M. Marra). Transaction Publishers (Piscataway, New Jersey).

MFA Denmark (2010) *Freedom from Poverty – Freedom to Change, Strategy for Denmark's Development Cooperation.* Ministry of Foreign Affairs of Denmark, Danida (Copenhagen).

MFA Finland (2009) *Evaluation of the Sustainability Dimension in Addressing Poverty Reduction (89886201): Annex B (Terms of Reference).* Ministry for Foreign Affairs of Finland, Office of the Under-Secretary of State, Evaluation of Development Cooperation (Helsinki).

MFA Finland (2011) *Evaluation of Country Programmes Between Finland and Nepal, Nicaragua and Tanzania (89889101): Annex B (Terms of Reference).* Ministry for Foreign Affairs of Finland, Office of the Under-Secretary of State, Evaluation of Development Cooperation (Helsinki).

MFA Netherlands (2007) *Our Common Concern: Investing in Development in a Changing World. Policy Note Dutch Development Cooperation 2007–2011.* Ministry of Foreign

Affairs of the Netherlands, Development Cooperation (The Hague).
MFA Netherlands (2009) *Results in Development, Report 2007–2008*. Ministry of Foreign Affairs of the Netherlands, Development Cooperation (The Hague).
MFA Norway (2009) *Climate, Conflict and Capital, Norwegian development policy adapting to change. Report No. 13 (2008–2009) to the Storting*. Ministry of Foreign Affairs of Norway (Oslo).
MFA Sweden (2003) *Shared Responsibility: Sweden's Policy for Global Development. Government Bill 2002/03: 122*. Ministry for Foreign Affairs of Sweden (Stockholm).
MFA Sweden (2008) *Global Challenges – Our Responsibility: Communication on Sweden's Policy for Global Development. Government Communication 2007/08: 89*. Ministry for Foreign Affairs of Sweden (Stockholm).
Norad (2010) *Evaluation of Norwegian Development Cooperation 2009*. Norad, Ministry of Foreign Affairs of Norway (Oslo).
OECD (2014) *Net Official Development Assistance from DAC and Other Donors in 2013*. www.oecd.org/dac/stats/documentupload/ODA%202013%20Tables%20and%20Charts%20En.pdf (consulted 13 Feb 2016).
Ritchie, J., Lewis, J., Nicholls, C.M. & Ormston, R. (editors, 2014) *Qualitative Research Practice: a Guide for Social Science Students and Researchers, Second Edition*. SAGE (London).
Robson, C. (1993) *Real World Research: a Resource for Social Scientists and Practitioner-Researchers*. Blackwell (Oxford).
SDC & SECO (2013) *Report on Effectiveness of the Swiss International Cooperation in Climate Change Mitigation and Adaptation Interventions 2000–2012, including Report on Climate 2011–2012 from 0.5% ODA-Bill (Tender Document)*. Swiss Agency for Development and Cooperation (SDC) and State Secretariat for Economic Affairs (SECO) (Bern).
SDC & SECO (2014) *Public Report on Effectiveness 2014 Swiss International Cooperation in Climate Change 2000–2012*. Swiss Agency for Development and Cooperation (SDC) and State Secretariat for Economic Affairs (SECO) (Bern). www.seco-cooperation.admin.ch/themen/01033/01130/05122/index.html?lang=en (consulted 20 Feb 2016).
UK (2002) *International Development Act 2002*. www.legislation.gov.uk/ukpga/2002/1/contents (consulted 12 Sep 2016).
UNFCCC (2016) *Summary of the Paris Agreement*. United Nations Framework Convention on Climate Change eHandbook. http://bigpicture.unfccc.int/#content-the-paris-agreemen (consulted 12 Sep 2016).
World Bank (2010) *Phase II: the Challenge of Low-Carbon Development: Climate Change and the World Bank Group*. Independent Evaluation Group, The World Bank (Washington, DC).
World Bank (2015) *World Development Indicators*. http://data.worldbank.org/data-catalog/world-development-indicators (consulted 13 Feb 2016).

Chapter 2

Core evaluation criteria

Criteria as points of view

This book's first area of focus lies in how to assess the performance of aid investments regardless of their specific purpose, by asking questions such as the following.

1 Do they make sense in terms of policy (i.e. the prior decisions that the collaborators have made) and context (e.g. the distribution of wealth and power, and of the likely costs and benefits of the investment, within the societies involved)?
2 Are they a defensible use of public funds, relative to alternative uses of the same money, including spending nothing?
3 Are they being managed tolerably well, without excessive conflict, incompetence, and corruption?
4 Are they inducing (i.e. facilitating, encouraging, enabling, buying) progress towards whatever their specific aims might be?
5 Are they having effects broader, more significant, and less predictable than the designers of the investments originally intended?
6 Are they creating a momentum towards irreversible change, for good or ill (and how to tell which)?
7 Is whatever they have achieved, are achieving, or might achieve, vulnerable or resilient to factors beyond their control (i.e. might some socio-political, economic, or environmental 'wild card' undermine or obliterate any gains, or make new kinds of gains possible)?
8 Are the various actors involved in the investment working together to harvest synergies, or at cross-purposes to cancel out each other's successes (if any)?
9 Have lessons been learned, or general principles and their practical applications tested and clarified, that can be used to improve the design of investments in the future or in other locations?

Answering questions like these is what the evaluation of aid performance is all about, although they are usually posed in terms of eight evaluation criteria, namely:

- *relevance* (question 1);
- *efficiency* (questions 2 and 3);
- *effectiveness* (question 4);
- *impact* (question 5);
- *sustainability* (question 6);
- *connectedness*, which may also be called 'risk' (question 7);
- *coherence*, which may also be called 'compatibility', 'coordination' or 'complementarity' (question 8); and
- *replicability* (question 9).

At their clearest and simplest, the terms of reference (ToR) of an evaluation will ask the evaluators to make an assessment of the investment against some or all of these eight criteria. The first five are most often used, since they have been defined by OECD/DAC (1991, 2010) and adopted as standard by the EU. But the officials and their advisers who write an evaluation's ToR may devise other criteria (such as the 'added value' of their own efforts), or use their own interests (which may be academic, personal, bureaucratic, or political) to devise lists of questions for the evaluators to answer which distract attention from the criteria themselves, or opt for idiosyncratic nuances according to the whims of their ministers, the vagaries of translations from their mother tongues to English, and the interests of the institutions for which they work, or the parliamentary committees to which they report. In other words, the purity of the criteria-based approach to evaluation is often obscured by the inventiveness of those who commission evaluations. One aim of this book is to make a case for a return to clarity on this issue. Thus, my view is that the DAC was on the right track in the 1990s, and that later efforts to make the evaluation process more sophisticated have resulted only in chaos, while also detaching the investments being evaluated from their real ecological and social contexts and consequences. Hence what is needed is to retain the DAC approach, broaden its list of core criteria, and revert to the use of common criteria and common sense in all aid evaluations.

Scoring performance by criterion

Investments supported by aid spending have to be evaluated by somebody, usually several somebodies with diverse backgrounds, languages, and cultures, and often several times during its life – by different teams, often the cheapest of those who bid for the jobs in each case. Such evaluations may be:

- at *pre-appraisal or appraisal* stage, when the original design – which is itself subject to questioning against the same evaluation criteria, or ought to be – is tested and revised on behalf of the donors and their collaborators;
- at the *inception* stage, when the personnel responsible for implementation think through how the project can be delivered in practice;

- at the *mid-term* stage, when strengths and weaknesses are identified and recommendations made on whether to change, extend, slow down, or speed up the project, to increase, reduce, or replace its staff, or to review its management systems; and/or
- at the *post-completion* stage, when lessons are learned from particular investments.

However, 'mid-term' reviews can be repeated every 2–3 years in a long project, or delayed until they merge into final evaluations in a short one. In addition, *synthesis evaluations* may be done, when multiple projects with similar themes – or even entire portfolios of hundreds of projects over many years – are compared, and useful patterns are sought. Finally, *meta evaluations* are also sometimes commissioned in which the quality of previous reviews is assessed, to shed light on whether or not the information gained from all the other evaluations is reliable and/or useful.

This diversity of participants and purposes, multiplied by an inconsistent vocabulary and the complex nature of the investments concerned and the environments where they occur, has eroded comparability between the findings of different evaluations. This applies equally to studies of investments with similar themes in different times and places, and to successive studies of the same investments over time. Without a common yardstick against which to report progress and failure, or strengths and weaknesses, progress cannot be judged and nothing can be compared easily against anything else. To avoid this outcome, in the DAC approach any investment could be examined from the point of view of 'relevance', for example, and judged transparently on whether and to what extent it responded to need and policy. Reasons for, or evidence in support of, a final judgement could be assembled, referenced and explained, and even where total compliance with stated needs and policies could be demonstrated, any deficiencies in those underlying assumptions (i.e. whether real needs have been identified, and appropriate policies articulated) could be discussed. Relevance thus became a useful and consistent point of view from which to examine any investment at any time. By adding a similar treatment using the other core criteria (effectiveness, impact, etc.), a fairly complete picture of the investment could be built up in the form of short analyses from each point of view, providing a meaningful assessment of performance presented *in the same terms* as any other assessment of performance for any other investment. In my view, this is a perfectly valid way to detect meaningful patterns within a universe of ODA investments that would otherwise generate an infinite amount of meaningless information. While the DAC strategy can be improved upon, it emphatically should not be abandoned in favour of an *ad hoc* approach driven by passing political, bureaucratic, and academic enthusiasms.

There are two ways to build on and improve the DAC approach, one fundamental and one analytic or presentational. The first involves asking whether the original five core criteria identified in the 1990s are sufficient to meet our collective needs. My answer is that they are not, and that three additional core criteria are needed, to capture the issues of 'connectedness' (primarily to address the phenomena of climate change and globalisation), 'coherence' (primarily to address the unnecessarily wasteful and contradictory practices of aid agencies), and 'replicability' (primarily to address the lack of systematic learning from aid spending). Meanwhile, to build on the presentational system, a way is needed to 'short-hand' the innumerable essays on the core evaluation criteria generated by evaluating thousands of ODA investments consistently over time, which calls for a scoring system that can make it easier to spot important patterns of strength and weakness within ODA portfolios.

One proven candidate for such a scoring system derives from EC monitoring practice. In this, a score of 'a' means *very good* (i.e. no real need for improvement was detected); 'b' means *good* (i.e. a few points required improvement, but the activity was otherwise sound); 'c' means *some problems* (i.e. significant improvement should have been required); and 'd' means *serious deficiencies* (i.e. the action should have been re-thought or should not have been supported). With intermediates ('a/b', 'b/c' and 'c/d'), this system creates seven possible scores, and they can be transformed for analytical purposes into numerical equivalents (d = 1, c/d = 2, c = 3, ... a = 7). When applied to the core evaluation criteria, this allows aggregate performance statements to be made and scores compared objectively among numerous investments, highlighting patterns of performance so that reasons for these patterns can then be explored. In large teams and with numerous investments to be evaluated, responsibilities will often need to be divided between team members, meaning that all have to share a common understanding of the criteria, the nature of relevant evidence, and the scoring system. It has been found that this can be achieved with a few hours' training, plus simultaneous trial assessments of the same intervention by the co-worker and team leader, and discussion of the results.

The key features of this approach are:

- that the performance of every investment is examined from the points of view of eight consistently defined core evaluation criteria (the DAC five, plus three new ones);
- an evidence-based, transparent, and defensible judgement is made on the investment's performance against all eight criteria; and
- a score is presented that is rooted in the evidence assembled, and can be used to draw attention to strengths and weaknesses within investments (either instantaneously or to reflect changes over time) and across portfolios of investments, reasons for which and lessons from which can then be explored.

In practical terms, where an aid portfolio comprising hundreds of projects is assessed for effectiveness, say, and half of them are found to be only moderately effective or worse, weeding out the worst performers and applying lessons from the others will raise the average score of the portfolio as a whole. This kind of macro-level approach can be helpful, *provided that all findings are rooted in a systemic understanding of each project*, in motivating and targeting reforms that will tend to produce an increase in mean score across multiple core evaluation criteria. This systematic process will be critical to maximising the performance of aid spending on the climate change response in coming years. To lay the groundwork for it, the following paragraphs briefly explain the eight core evaluation criteria and illustrate the assessment of particular investments from each point of view. After that, I will present a more detailed survey of the application of this method to the sample of aid investments listed in Appendix 1.

The eight core criteria

Relevance

The assessment of relevance focuses on whether and to what extent the investment responds to the needs of the intended beneficiaries in their political, economic, and ecological contexts, and is aligned with the overall policy environment through a convincing theory of change (Boxes 2.1 and 2.2). Judgements are based on information that is usually contained in the sections on analysis, background, justification, and rationale of documents dealing with investment design, and the same preliminary sections of plans, reports, and evaluations of implementation and results. Signs to be looked for include those of attention being paid to the analysis of key issues and application of key policies that help the investment contribute to sustainable development aims (e.g. those that promote local control of ecosystems, locally accountable governance, public understanding of environmental and livelihood links, local business skills, conflict management arrangements, and attention to other cross-cutting themes).

Box 2.1 Relevance – a high score in Nepal

Finnish support to the peace process and good governance. "The absence of widespread political violence is a pre-condition for overcoming all other development challenges. Lasting peace cannot be achieved, however, without also promoting good governance (in our definition: *stable, lawful and effective governance maintained by accountability to an informed electorate*), since this is a precondition for sustainably reducing corruption and inequality, which are among the root causes of conflict. Recognising this, and also that Finnish resources alone are insufficient to the task, Finnish efforts join with those of other donors in actions directed to calming conflict (i.e. creating the space for dialogue, negotiation and compromise), building negotiated agreements on the ground-rules of governance (i.e. creating an agreed constitution), and promoting mutual respect and reducing oppression, exploitation and suffering (i.e. protecting human rights). These efforts have taken the form of: (a) supporting the Nepal Peace Trust Fund, which provides resources for managing camps, reintegrating former combatants, rehabilitating displaced people, organising elections, strengthening the rule of law, and supporting the peace process; (b) supporting constitution building, mainly through the International Institute for Democracy and Electoral Assistance (IDEA), which promotes the flow of knowledge into and among the factions that are trying to hammer out a viable constitution; (c) supporting the Office of the High Commissioner for Human Rights, which addresses discrimination and human rights, the strengthening of human rights institutions and civil society, and accountability, impunity and the rule of law; (d) supporting the National Human Rights Commission, which investigates and seeks to repair human rights violations of all kinds; and (e) participating in the Utstein Group, which promotes the mainstreaming of a peace-building and peace-sensitive approach in all bilateral ODA and non-governmental organisation (NGO) interventions. This is all supported by complementary investment through Finnish NGOs and their Nepalese counterparts, with particular attention to poor districts, marginalised and excluded groups, and the themes of peace-building and conflict transformation" (Caldecott, Hawkes, Bajracharya & Lounela, 2012: 73–74).

Source: Appendix 1 (33).

Box 2.2 Relevance – a low score in Finnish academia

Finnish support to development research. "None of the main modalities for supporting development research (MFA-commissioned, Academy of Finland/AoF-led, and contributions to international institutions) are described as more than broadly relevant to the general theme of understanding situations in developing countries, let alone to enhancing the design of anti-poverty and pro-sustainable development efforts. Little new knowledge is generated that would help MFA do its job more effectively, opportunities are not being exploited for research in partner countries to reinforce and amplify Finland's engagement with those partners, and there is little sign of the research cooperation that might otherwise build capacity in developing countries. There are moreover signs that the Finnish research community actively resists moves to encourage greater relevance, as for example, the AoF's Development Research Strategy which is explicit 'that moves to strengthen policy relevance and applicability of research must not compromise the autonomy of research and the freedom of the researcher' (page 44), while 'the Academy is quite insistent that it will best serve the research interests of the university community by emphasising academic quality over policy relevance' (page 102)" (Caldecott *et al.*, 2010: 233).

Source: Appendix 1 (18).

Efficiency

The assessment of efficiency focuses on whether and to what extent the investment contains measures that through elegance and accountability promote sound management and value for money, yielding confidence through counterfactual ('what if not') and other analysis that the same or better results are unlikely to be achieved through different means, or with lower overall expenditure, or with different rates of expenditure, or with no investment at all (Boxes 2.3 and 2.4). Judgements are made by examining practical issues such as the quality of leadership and the appropriate expression of roles, rights, and responsibilities among partners and stakeholders, as well as flexibility, accountability, and the use of opportunities for synergy among development actions and actors. Signs to be looked for include management and cost issues raised in progress and evaluation reports.

Box 2.3 Efficiency – a high score in China

Mid-Term Evaluation of the EU–China Biodiversity Programme (ECBP), Visibility and Awareness Component. "Outputs are diverse, abundant and of high quality, especially considering the relatively limited resources (€2 million routed through a European for-profit consulting firm). The Visibility and Awareness Component is also responsive to the EU Delegation's wish that credit for the ECBP is given appropriately to the programme's financiers rather than just to its implementers" (Caldecott *et al.*, 2009: 21).

Source: Appendix 1 (9).

Box 2.4 Efficiency – a low score in South-east Asia

Mid-Term Review of the Association of Southeast Asian Nations (ASEAN) Regional Centre for Biodiversity Conservation. "The project was plagued by delays in releases of funds, which caused a tremendous set back in operational performance over the first half of the project life. These resulted from a combination of bureaucratic procedures imposed by the EC,

EC Delegation and Government of the Philippines (GoP), compounded by poor communication and a lack of trust between EC, GoP and the first two EU finance and administration advisers. The latter were perceived as not qualified to establish efficient and effective financial and administrative systems for the project, in line with accepted accounting practices. The processes of removing and replacing these advisers, added to other lengthy institutional response, referral and approval processes, severely affected project efficiency. With two years virtually lost, the overall efficiency of a five-year project cannot really be seen as satisfactory, even though great improvements have since been made" (Caldecott *et al.*, 2002: 30).

Source: Appendix 1 (2).

Effectiveness

The assessment of effectiveness focuses on whether and to what extent results are being achieved or anticipated that contribute to progress towards achieving the purpose of the investment, which are usually specified in its own terms but may include contributions to broader cross-cutting objectives and other policy goals (Boxes 2.5 and 2.6). Such results may take the form of changes in relationships, institutions, legislation, administration, processes of dialogue and conflict resolution, and quality of life indicators that can be reliably attributed to the investment. Judgements are therefore made with respect to such changes, and signs to be looked for include relevant descriptions, analyses and data contained in progress reports, evaluations, and commentaries by independent observers. Since effectiveness is usually of particular public interest, evidence for it is especially important and may take the following forms.

- In a climate change mitigation investment, direct evidence might include data on real GHG emission reductions (or proxies on energy efficiency), provided that some quantified baseline exists and some reasonable protocol to describe measured changes was applied. If no such data, baselines, protocols, or measurements exist in the project documents, the reviewer should say so.
- In a climate change adaptation investment, direct evidence might include documentation and/or witness statements to the effect that environmental events and changes that are believed to be linked to climate change (e.g. droughts, fires, floods, sea-borne storms, dust-storms, cold snaps, heat-waves, or creeping salt-water intrusion) are being coped with (in any sense – including social, financial, environmental, and political resilience, and early warning) better after the project than before. If no such evidence exists in the project documents, the reviewer should say so.
- Indirect evidence may also be used, being any other information possessed by the reviewer that is relevant to forming a judgement on the likely

effectiveness of the investment. This might come from other, similar projects that they know about, or from transparently applying any other kind of inference to build a case for or against likely effectiveness of this particular project.

Box 2.5 Effectiveness – a high score in Mongolia

Sustainable Land Management for Combating Desertification. "Community-based organisations in all project target areas were encouraged and enabled to practise the fencing of hay fields, to the extent of 1–4 ha of pasture each year in each place. Fenced grassland recovers well from over-grazing and vegetation cover and plant yields were shown to increase dramatically, providing an emergency grazing reserve for livestock in harsh winters, a rehabilitation area for ill or weak animals, and a surplus of hay for sale. This technique is effective at increasing ecological resilience and is being replicated. Rehabilitating and protecting springs, wells and catchments has been shown to improve water supply and security, while planting of native trees and establishing mechanical and biological barriers have been shown to reduce wind erosion and sand movements. The project also piloted the use of large numbers of fuel-efficient stoves (provided by a German aid project) with fuel savings of 40–50 per cent, thus reducing pressure on woody vegetation. The project has successfully introduced a wide range of 'no regrets' measures to build social capital (e.g. by organising 109 herder groups and 13 forest user groups involving more than 1,200 households), to build public knowledge on sustainable land management (SLM, e.g. by delivering 74 training sessions involving over 8,500 participants, 53 per cent of them female, focused on pasture management, traditional rotational grazing practices and the planting of trees and shrubs, and by promoting environmental education at scores of schools), to improve local land management planning (e.g. by training land managers in planning and mapping), and to contribute to the further development and institutional coordination of national policy and legislation (e.g. revision of the National Action Plan for Combating Desertification, capacity building at the National Committee for Combating Desertification, drafting of pastureland laws, development of university curricula). These measures are considered very likely to contribute significantly to the project's three intended outcomes: (a) strengthened coordination mechanisms, institutional and human resources capacity and knowledge base to promote SLM and desertification control; (b) SLM mainstreamed into national, provincial and local policies, strategies and regulatory framework; and (c) pilot testing, demonstrations and scaling-up community based approaches in integrated natural resources management with a focus on grassland and water management and sylvopastoralism (i.e. the grazing of livestock and growing of trees on the same piece of land). These outcomes in turn are consistent with almost everything known about how to increase the security and resilience of human and ecological systems to the effects of climate change in an environment like Mongolia's, at least at the level at which it aggravates existing trends that have multiple causes (in this case an explosive growth in herding families and livestock numbers and a prevalent open-access grazing regime), and to an extent also at the level of disastrous winter weather" (Caldecott *et al.*, 2014: 166–167, project SDC 7F–06465).

Source: Appendix 1 (40).

Box 2.6 Effectiveness – a low score in Vietnam

Mekong Market Development Portfolio Project. "The project targeted GHG emission reductions through promoting sustainable natural resources management in the bamboo sector. However, from the project documentation it is not possible to define to what extent the project managed to contribute to the sustainability of natural resources management. Business development activities were implemented, but it is unclear how large an area of land has been managed better than before. Such a project is likely to have some small relevance both to mitigation (through accumulation of woody biomass) and mitigation (through re-vegetation of degraded slopes). According to the credit proposal, sustainable natural resources management was a core dimension of the project's analytical and implementation work. However, according to the project logframe, there were no activities that would have directly contributed to the quality of natural resources management. An environmental impact assessment of bamboo cultivation was to be done and market development activities promoted to increase income opportunities from bamboo production, but it is unclear if GHG emission reductions were obtained through natural resources management activities during the project or not. The programme review credited the project with developing high quality investment support mechanisms for bamboo processors. These mechanisms included Sustainable Supply Services where the sustainability of bamboo supplies was to be secured. Business partner demonstration was established with Tien Dong company with 2750 ha of bamboo plantations. Unfortunately the project and the lead partner Prosperity Initiative Community Interest Company, which was responsible for the investment support mechanisms, were run down before the services became widely available. Other general market development activities of the project may have contributed to mitigation through promoting sustainable natural resources management, which might also have contributed to improving adaptation capacity in rural areas. These potential impacts cannot be demonstrated, however, due to lack of information on the quality and quantity of actual improvements in natural resources management created through project activities" (Caldecott *et al.*, 2014: 301, project SDC 7F–05697).

Sources: Appendix 1 (40, 46).

Impact

The assessment of impact focuses on whether and to what extent the investment as a whole is likely to exert influences that are wider and longer-term than its results considered alone (Boxes 2.7 and 2.8). Using the impact criterion thus draws attention to strategic changes attributable directly or indirectly to the intervention, which may be more or less subtle in nature. Impacts might include those in which the investment directly or indirectly contributes to enhanced well-being, for example through changes in skills, education, relationships, institutions, legislation, and administration. Negative impacts should also be considered, and could include unintended economic externalities, perverse incentives, population movements, and ecological deterioration. Judgements are made with respect to such changes, and signs to be looked for are relevant descriptions, analyses and data contained in strategic national or regional evaluations, and commentaries by informed and independent observers. Impact and sustainability can be hard to consider separately, since the various attributable consequences of an investment during and immediately following its deployment (i.e. 'impact') may exert continuing influence thereafter, either in themselves or as a result of changes that they have induced (i.e. 'sustainability').

Box 2.7 Impact – a high score in Nepal

Finnish country programme. "Although impact is hard to attribute to individual participants in multi-actor interventions, very important impacts at national level will flow from investments in peace-building, governance and human rights and in the education sector. Significant impacts on health, livelihoods and equity at community and to some extent district level are being achieved by the bilateral water, sanitation and hygiene projects, which have established best practices that are being used by government and other donors, so are also having an impact at national level. Finnish support to the Forest Resource Assessment and other inventory, survey, mapping and capacity-building activities in forestry and land use help create opportunities for impact if they are used by other actors in applying forest management systems, in developing REDD [reducing (greenhouse gas) emissions from deforestation and (forest) degradation] activities, in effective planning and in controlling corruption. Climate change early warning systems can have important impacts if information is disseminated and used effectively. Meanwhile, the environment projects are likely to have substantial impacts, owing to the long-term health benefits of avoided pollution" (Caldecott, Hawkes, Bajracharya & Lounela, 2012: 77).

Source: Appendix 1 (33).

Box 2.8 Impact – a low score in Tanzania

Finnish participation in the Development Partners Group (DPG). "One area where Finnish expectations were certainly over-ambitious in terms of impact was the policy leverage that could be exerted on government through dialogue forums linked to the DPG working groups. For example, by co-chairing the thematic groups on natural resources and tourism as well as on environment, Finland with a few other donors have managed to keep cooperation alive in the forest sector, but this effort has not been reflected in government plans and budgets where the forest sector, despite its economic value and potential, is not a high priority. Donors' attempts to steer policies through the dialogue structures have sometimes been counter-productive, when government representatives have felt that donors have tried to use technical discussions to apply pressure on sovereign matters" (Caldecott, Valjas *et al.*, 2012: 84–85).

Source: Appendix 1 (35).

Sustainability

The assessment of sustainability focuses on whether and to what extent the investment is likely to have effects that will continue because of irreversible changes that have been induced, often in laws, institutions, relationships, and public opinion (Boxes 2.9 and 2.10). Using the sustainability criterion thus draws attention to whether the intervention is likely to induce beneficial results that continue indefinitely. Signs to look for include:

- changes to policies, laws and regulations, systems and working practices, establishment of new forums, creation of new permanent staff positions;
- changes to fiscal arrangements and budget allocations, creation of thriving businesses with local participation in benefits;

- changes to trends in environmental deterioration and ecosystem restoration, introduction of incentives and resource management systems that reward sustainable use of ecosystems; and
- introduction of new ideas, groups and activities that contribute to environmental protection.

Box 2.9 Sustainability – a high score in Thailand

Final Evaluation of the Coastal Habitats and Resources Management (CHARM) Project. "Key sources of sustainability are expected to lie in the continued commitment of the Department of Fisheries (DoF) and the Department of Marine and Coastal Resources (DMCR) to invest further in the co-management approach, in the ongoing implementation of the Thai decentralisation process with the adoption of the new National Constitution, and in the National Integrated Coastal Management Dialogue that the project has initiated with, amongst others, the Chonburi Network, Andaman Triangle and Trang Seas initiatives, as well as the continued participation of NGOs. The Andaman Triangle Network is an inclusive forum that has met ten times under CHARM sponsorship, and most recently an 11th time with the support of DMCR. It appears to be well appreciated by NGOs, communities and government, and seems very likely to continue in operation. Meanwhile, the monitoring, control and surveillance network established by CHARM appears well grounded and widely supported, and was often described as being one of the strongest aspects of the CHARM project. At a community level, the project has facilitated the development of 69 conservation groups, 49 aquaculture groups and 35 fishing/marine groups. When asked by Wetlands International how CHARM had helped their livelihoods, 'just under half mentioned the more careful use of coastal resources, and development of alternative livelihood occupations to increase income.' Finally, beyond the institutional level, individual experiences on co-management will likely be carried forward, using the knowledge and skills developed by NGOs, project and DoF staff, and others at the Tambon (sub-district) level" (Caldecott *et al.*, 2007: 26–27).

Source: Appendix 1 (4).

Box 2.10 Sustainability – a low score in Tanzania

Finnish country programme. "Debt forgiveness and general budget support were always intended to be temporary measures which awaited a more favourable fiscal situation, one that is now arising in Tanzania as the government increases collection of domestic tax and other revenues. As temporary measures, they were not intended to be themselves sustainable, but they have made possible improvements in economic performance which will have had a sustainable effect to the extent that these benefit the poor and continue to do so. In other areas, however, sustainability is a concern. In forestry and agriculture there are outstanding issues related to land ownership which affect negatively the sustainability of results. Moreover, because of weak national capacity, the tendency has been to opt for separate implementation structures rather than to institutionalise interventions within government structures. This has a detrimental effect on sustainability in the absence of effective exit strategies (e.g. in the cases of the National Forest Resources Monitoring and Assessment and Sustainable Management of Land and Environment projects). A similar move from institutionalised interventions to separate implementation structures has not yet taken place in the Local Government Reform Programme, the result being that the latter is so weakly managed that some donors are considering withdrawing from it (as others have already done). The sustainability of projects in the so-called 'new areas of cooperation' is weak (these being projects initiated by ministerial fiat in late 2007, comprising investments in information and communication technology, the energy sector, a Sustainable Development Institute, and a number of smaller initiatives on UN support, refugees, seed potatoes and geological mapping), whether institutionally, or financially and politically" (Caldecott, Valjas *et al.*, 2012: 85).

Source: Appendix 1 (35).

Connectedness

The assessment of connectedness focuses on whether and to what extent the investment is likely to be resilient to external factors and influences over which it has little or no control, and contains measures by which these influences are to be mitigated (Boxes 2.11, 2.12, and 2.13). Using the connectedness criterion thus draws attention to the extent to which targeted systems are connected to others that may affect outcomes, for example by undermining planned development efforts or stranding invested assets (see Chapter 4). Judgements are based on whether there are signs of plausible analysis of systems and factors outside or upstream of the area of cooperation that have the capacity to jeopardise it, and/or whether adequate mitigation measures are included, including the issue of whether and how global challenges (e.g. climate change, increasing disaster risk, growing competition over natural resources, food and water) feature in the cooperation pathway from policy dialogue to implementation. Two examples each of 'high', 'medium' and 'low' scores are given to highlight the inherent complexity of judgements concerning the connectedness criterion.

Box 2.11 Connectedness – high scores in a global programme and in Mozambique

Global. "The reasons behind the strong effectiveness in parts of the portfolio are diverse but a number of common features were detected, [including] where opportunities to participate in disaster-oriented insurance coverage were systematically broadened, especially where disaster risk and adaptation strategies were integrated and initiatives screened for climate vulnerabilities and proofed against them, [and] where core funding was provided for an institution known or reputed to have a substantial record of service delivery relevant to climate change mitigation or adaptation, or both, and with global or regional reach and the capacity to distribute lessons learned widely" (Caldecott *et al.*, 2014: 51–52).

Mozambique. The city of Beira is the second largest city in Mozambique and its port and railway connections are the principal source of wealth and development for the city and its hinterland in Mozambique, Zimbabwe, Zambia and Malawi. Beira is situated in a very flat and low sedimentary area, and the *Study on the Impact of Climate Change on Disaster Risk in Mozambique*, produced by the National institute for Disaster Management in 2009, identified the Beira coastline as highly vulnerable to the effects of climate change and calls for protective measures along with the installation of monitoring and early warning systems. The overall goal of the intervention is to safeguard the city's development potential and protect the citizens of Beira against the effects of climate change through preventive measures against rising sea levels and more frequent and stronger cyclones. A complementary investment in coastal protection is thus also a way to safeguard past and future investments in the institutional development and service delivery capacity of the municipality (Caldecott *et al.*, 2014: 360–362, project 7F–07923).

Sources: Appendix 1 (41, 45).

Box 2.12 Connectedness – medium scores in Tanzania and Nepal

Tanzania. "The country programme is fragmented and therefore diverse, with each part having its own strengths and vulnerabilities, and there are few common connectedness issues. Moreover, the programme is embedded within a vast structure of aid activities created by dozens of donors in partnership with government over many years. The only externalities that could realistically threaten the whole edifice are massive political turmoil (which is perhaps least likely in modern-day Tanzania than in almost any other African country), wholesale collapse of international financial systems, and climate change. Only the latter vulnerability is within the scope of the Tanzanian development process to affect, which it could do through measures to promote adaptation (e.g. through climate-proofing, ecosystem management and disaster risk reduction) and mitigation (e.g. through REDD and renewable energy), although unlike some donors (e.g. Norway, the UK) Finland has taken only small (if strategic) positions on these subjects, and the government seems little interested in them at present" (Caldecott, Valjas *et al.*, 2012: 88).

Nepal. "At a country programme level, the target systems are those that define and maintain the country's economic performance and the quality of its prevailing social relationships. These systems in Nepal are vulnerable to political conflict (e.g. between classes, parties, secessionist movements), external interference (e.g. by powerful neighbours advancing their own political, trade, cultural or territorial interests), external economic events (e.g. recession affecting remittance payments) and physical processes (e.g. climate change, extreme weather, seismic shock). Nepal is an ecologically fragile country emerging from a long civil war with deep internal divisions, uncertain political institutions and two populous, competitive and assertive neighbours. It is therefore inherently vulnerable to connectedness issues, but the country programme itself is exemplary in addressing at least some of the root causes of this vulnerability. By targeting the causes of internal conflict (e.g. inequality, exclusion, ignorance, lack of solidarity), promoting dialogue-based rather than violence-based solutions (e.g. through peace-building, human rights and constitutional development), and laying the foundations for development and greater economic security (e.g. through efforts on trade and investment, resource inventories, vocational training, land tenure and community-based resource management), the programme is contributing to creating a stronger country that will be more resilient to future shocks beyond its control. Connectedness and vulnerability issues remain, and there is room for improvement in addressing them (particularly on the climate change side), but many current signs are reasonably positive" (Caldecott, Hawkes, Bajracharya & Lounela, 2012: 79–80).

Sources: Appendix 1 (35, 33).

Box 2.13 Connectedness – low scores in Namibia and the Kyrgyz Republic

Finnish aid in Namibia. "Namibia is facing numerous environmental challenges, including water scarcity and pollution, land degradation and deforestation, loss of biodiversity and risks of natural disaster. [The] climate change situation is likely to further aggravate environmental deterioration, particularly related to water scarcity and land degradation (increased erosion from over grazing or intense occasional rains), affecting health and food production" (Caldecott *et al.*, 2010: 182).

Finnish aid in Central Asia. "The Lung Health Project in the Kyrgyz Republic focuses on improving the diagnosis of lung diseases, but cannot reduce their incidence since it is not coupled with any measures to reduce smoking or other known causes of lung pathology. The absence of competent departments of preventive medicine, public health and health education in the Kyrgyz Republic, and an inability to address the issue of lung health in its entirety, are factors beyond the reach of Finnish development cooperation alone. The primacy of national sovereignty and state-building in post-colonial situations may limit the effectiveness of Finland's seemingly justified involvement in region-wide projects on collapsing water supply, border issues, out-migration of labour, energy, narcotics and organised crime, and religious extremism. Finland's support for the Central Asia Regional Information and Cooperation Centre is an example of an activity being crippled by inadequate coordination and ineffective cooperation among area governments" (Caldecott *et al.*, 2010: 228).

Sources: Appendix 1 (12, 17).

Coherence

The assessment of coherence focuses on whether and to what extent the investment is likely to maximise opportunities for synergy with, and mitigate interference from, the plans and actions of other actors (Boxes 2.14 and 2.15). Factors include: *compatibility* (i.e. how well the goals of all participants are taken into account and where necessary reconciled); *coordination* (i.e. the existence and likely use of forums to sustain dialogue among stakeholders); and *complementarity* (i.e. how well participants' policies, plans, actions and choices support one another, and the degree of harmony among partners in achieving desired outcomes). Each of these have a different meaning depending on whether they focus at the level of tactics, strategies, policies, plans, dialogue or decision-making processes, and concern events within donor institutions or between those and the agencies of recipient governments. Using the coherence criterion thus draws attention to whether the policies, plans and activities of all development actors are in line with each other.

Box 2.14 Coherence – a high score in Nepal

Finnish country programme. "Because so much development cooperation in Nepal has a multi-donor aspect and is in line with plans developed by the ministries concerned, both donor-donor and donor-government coordination systems are important and prominent. Aside from day-to-day informal dialogue and joint problem solving, which is characteristic of relations between the Finnish Embassy, government and other donors, there are a number of formal coordination arrangements, including a joint programme on the peace process and sectoral coordination committees which are often led by relevant ministries, for example in the education sector where there is a multi-donor basket funding arrangement with nine donors jointly making funding decisions and with the Ministry of Education managing overall finances. Finland's development community is very diverse and comprises the Ministry for Foreign Affairs (MFA, with its 'troika' of ministers and related portfolios of diplomacy, trade, and development), partner country embassies, numerous governmental, non-profit/non-governmental, academic and technical institutions, all funded at least partly by government under various instruments, plus for-profit enterprises which accept government contracts or government facilitation of investment opportunities. While the individual entities each have their own specific purposes, the MFA and embassy seem able to maintain good communication with all concerned, and to orchestrate them with a light touch that allows them all, most of the time, to advance their own interests without competing with each other in an unhealthy way, while also advancing the interests of Nepal in line with its own and with Finland's development priorities. There are also sufficient lines of communication among development actors to ensure coherence within the sectors where Finland is active" (Caldecott, Hawkes, Bajracharya & Lounela, 2012: 78).

Source: Appendix 1 (33).

Box 2.15 Coherence – a low score in Kenya

Finnish country programme. In the Governance, Justice and Law and Order Sector programme "there were considerable problems in coordination among the development partners as well as between the Government of Kenya and the development partners", while the Programme for Agriculture and Livelihoods in Western Communities "failed to explore synergies with any of the several relevant donor programmes active in Busia County" (Uusihakala *et al.*, 2015: 20–21, 23–24).

Source: Appendix 1 (48).

Replicability

The assessment of replicability focuses on whether and to what extent the investment has the potential to yield lessons that can be used to improve actions in the future or elsewhere (Boxes 2.16 and 2.17). Using the replicability criterion thus draws attention to whether lessons have been learned so that interventions in the future or in other locations can be modelled on improved versions of past ones. Judgements are based on reasoned analysis to support or deny the expectation that improved programmes can be designed based on a previous set of choices (i.e. among aid modalities, sectoral interventions, etc.) that are expected to be effective against new but similar challenges, and which are connected through a common policy or planning approach.

Box 2.16 Replicability – a high score in Nicaragua

Finnish support to sector-wide programmes. "The four-pronged approach involving institutionalised sector-wide programmes, sectoral budget support, basket funds and pilot projects seems eminently replicable as well as likely to yield results that would score highly on other criteria (especially strategic effectiveness, impact, sustainability and coordination). Much would still depend on other factors and choices, however, including the selection of tools such as peer education that lend themselves to replication, the choice of themes that are particularly relevant to each country's particular circumstances, and investment in human resources" (Caldecott, van Sluijs *et al.*, 2012: 82).

Source: Appendix 1 (34).

Box 2.17 Replicability – a low score in Tanzania

Finnish country programme. "The use of general budget support is effective as a temporary way to advance the Millennium Development Goals, which is relevant to other developing countries in similar socio-economic situations. Other conclusions on replicability are rather negative, however, such as that a high price will have to be paid to correct over-ambitious expectations of policy leverage, the over-estimation of national programme management capacities, and the abandonment of evidence-based and collective decision making" (Caldecott, Valjas *et al.*, 2012: 91).

Source: Appendix 1 (35).

High-scoring investments

A pastiche of high scores from Boxes 2.1, 2.3, 2.5, 2.7, 2.9, 2.11, 2.14, and 2.16 simulates an impressive investment. Here, the collaborators would be working together in systematically targeting by far the most important issues, generating results that are cheap, diverse, abundant and of high quality, and strengthening capacity for sustainable resource use, equitable wealth creation, and livelihood resilience at all levels of society. Meanwhile, they would be creating new options by improving health, knowledge, participation, and transparency, reinforcing government support for improved, decentralised, and more responsive governance, anticipating and putting in place ways to resist negative cross-border influences, and consciously demonstrating and learning from systems and solutions in ways that can be used to improve many other investments.

Low-scoring investments

A pastiche of low scores from Boxes 2.2, 2.4, 2.6, 2.8, 2.10, 2.13, 2.15, and 2.17 simulates a less impressive scenario. Here, the collaborators would seldom speak to each other, their disparate activities would be weak and unfocused, and delayed by poor communication, distrust, and bureaucracy, while key stakeholders wandered off or went bankrupt due to lack of care. Nothing could be shown to be being achieved, and attempts at vigour would provoke hostile reactions, while deep and unaddressed flaws in such fundamentals as resource tenure, transparency, and institutional capacity would cripple any hope of sustainability. Anything that was achieved despite all this would be undermined by external factors that had neither been anticipated nor planned for. No one would be very interested in learning from the project, although some interested parties would nevertheless be holding it up as an example of best practice.

Real-life investments

Aid investments are seldom as good or as bad as these pastiches. Rather they have strengths and weaknesses, individually and collectively, as shown by an overview of the 2010 synthesis evaluation of the Finnish aid portfolio. Thus, "it was concluded that Finnish activities typically meet the priorities both of Finland and of her partner countries, respond to some but not all elements of the global development agenda, and usually satisfy the needs and wishes of cooperating partners. Partly as a result, Finns are widely respected for their skills, approach and attitudes. On the other hand, with some exceptions and despite often-strenuous Finnish efforts, donors and other stakeholders do not communicate well-enough with one another to achieve much synergy. Environmental and cross-cutting issues are weakly mainstreamed in Finnish activities, so overall sustainability is compromised. Moreover, to judge from the average scores and no doubt with many exceptions, the typical Finnish activity is not well-enough designed to be able to deliver much in the way of short-term results or longer-term and broader impacts, or to adapt to external pressures that may arise. Worst of all, the key aim of all these activities, poverty reduction, is vague, indistinctly measurable and little measured, so it is often assumed to be delivered even when it may not be, or when it occurs for other reasons" (Caldecott *et al.*, 2010: 24).

Other criteria and influences

Donor development policies

Much expert staff time is invested by donor agencies in getting their development policies 'just right', and in some cases assessing the effect of changing policies is built into the ToR of evaluation studies. For example, one focus of the 2011–2012 evaluation of the Finnish country programmes in Nepal, Nicaragua and Tanzania was on "how the Finnish development cooperation policy frameworks of 1998, 2001, 2004 and 2007 were transformed into practice" (Caldecott, Hawkes, Bajracharya, van Sluijs, Aguilar, Valjas, Killian & Lounela, 2012: 1). It was found that policy differences over time had a steering effect that by itself was quite weak, since they fused with other influences in exerting their effects. These influences included: opportunities identified by embassies to collaborate with other donors; needs and plans expressed by host governments; directions suggested by the Paris Declaration (e.g. in favour of sector-wide programmes); the desire to continue with long-term relationships (e.g. in particular regions of the country); the particular circumstances of the country (e.g. in emerging from civil conflict or from a prevailing political or economic system); and "the personal preferences of decision makers in the MFA, embassy or government" (*ibid.*: 8). A conclusion is that development policies must certainly be considered in evaluation work, but they are not always as potent in influencing events as their framers would like them to be.

A factor that affects the implementation of development policy is conflict with other policies or practices, in which, for example, agencies of a government may be investing in high-emission and biodiversity-depleting activities (e.g. oil-palm plantations in rain forests) while other agencies of the same government are simultaneously investing in low-carbon and biodiversity-friendly ones (e.g. REDD+ initiatives). Such cases result from the complexity and dynamism of the world economy interacting with patchy awareness of broader and longer-term consequences of investment decisions among diverse interest groups. They contribute to an area of overlap between coherence (avoided policy conflict) and connectedness (avoided vulnerability to external influences) that can make it hard to assess investments from either point of view. In these circumstances, an evaluator should follow and document each line of connection and conflict in order to minimise unexpected events and draw attention to issues that may undermine progress.

Partner satisfaction

The assessment of partner satisfaction focuses on whether and to what extent partners and stakeholders in an investment programme are satisfied with its processes and results, equally applied whether those partners are global, regional, national, local, or community institutions (Boxes 2.18 and 2.19). Using the partner satisfaction criterion thus draws attention to strengths and weaknesses in how a partnership is actually working, which ultimately depends on the perceptions of the partners. Judgements are therefore based primarily on interviews. Although all investments have multiple participants, most of whom must feel positive about the arrangement if it is to be viable and effective, the partner satisfaction criterion is particularly applicable where a long-term arrangement based on mutual benefit and co-adaptation is envisioned (see Chapter 7). As a criterion of performance it differs from the core eight, since it captures attitudes that may be positive or negative regardless of any rationally defensible point of view. Thus an investment might be very popular among stakeholders while being deeply flawed, or very unpopular despite it being near-perfect in all other respects. This is the realm of politics and public relations (and, often, personal or international relations, nationalism, or religion), so partner satisfaction should be studied and used cautiously, and mainly where it is the way in which a partnership is working that is the key issue.

Box 2.18 Partner satisfaction – a high score in academic cooperation

North–South–South (NSS) Higher Education Institution Network. In 2009, 34 higher education institution networks in 23 countries were being supported by Finland. "Finnish institutions consider the NSS Programme a valuable instrument to establish networks with their Southern counterparts, as it has been practically the only instrument for student and teacher exchanges from the South. Many Finnish exchange students considered the exposure to a different culture in a developing country a valuable experience, even if the academic gains remained limited. Also appreciated in the South were the opportunities provided by the NSS Programme for exposure to Finnish tertiary education and sectoral experiences (e.g. in the health sector) as well as for developing regional level collaboration, and the high-quality organisation and academic content of exchanges and intensive courses. Most respondents considered the relevance and contribution of the NSS Programme to their activities to be so high as to justify the burden of extra challenges and expenses they incur in its implementation. The performance by the Centre for International Mobility as programme manager was appreciated by both the MFA and the majority of Finnish network coordinators as well as their Southern counterparts" (Caldecott *et al.*, 2010: 254).

Source: Appendix 1 (20).

Box 2.19 Partner satisfaction – a low score in the energy sector

Finnish Support to the Energy Sector. "Government representatives met in country field visits noted slow administrative process and excessive interference from Helsinki as difficult to understand and inefficient at best. In both Kenya and Vietnam, seemingly arbitrary cancellations of funding were indicated as a major irritant and additionally affecting future partnerships and activities. In both countries, several knowledgeable stakeholders questioned whether Finland respected country ownership in project selection and implementation" (Caldecott *et al.*, 2010: 355).

Source: Appendix 1 (30).

Cross-cutting themes and mainstreaming

Some issues of grave importance to stakeholders in an investment are hard to score – either at all, because of a lack of data, or consistently, because of their subject matter – but must nevertheless be discussed in an evaluation. These include how the investment affects the interests of groups that are at risk of being excluded or systematically disadvantaged because of landlessness, caste, poverty, ethnicity, gender, age, faith or other reasons, or their correlates such as indigenous identity and forest dependence, any of which may also be associated with relative disadvantage between countries and sub-national regions such as districts and provinces, or urban and rural areas. They also include the prevalence of widespread diseases which may debilitate or kill among populations in ways that undermine the performance of aid investments. These are all often considered as 'cross-cutting themes' (CCTs), where these comprise the amorphous factors that should be considered by everyone in planning or evaluating aid investments. It can be hard to turn any of them into 'hard' procedures that make a real difference to processes and beneficiaries, but this is not impossible as shown by the best practice demonstrated by a Finnish project in Nepal (Box 2.20).

Box 2.20 Mainstreaming gender equity and social inclusion (GESI) in Nepal

"The Rural Village Water Resources Management Project (RVWRMP) focuses on some of the poorest parts of Nepal, ten mountainous districts in the Mid- and Far-Western development regions, with a total population of around 1.5 million. The government is barely present in the remotest areas: schools have no teachers, health posts have no medicines, and Village Development Committee secretaries live in district headquarters.

Malnutrition is common, particularly among children, and water-borne diseases are frequent. Men migrate to India in search of work for an average of eight months each year, and return with their meagre savings and, often, sexually transmitted diseases. Delivery of development services is minimal, and there is pervasive corruption and discrimination based on caste and gender. Girls are often denied education and married off as children, while women are commonly barred from their homes while menstruating and generally denied the right to participate in development processes of all kinds.

"A GESI strategy was developed for the project during its inception phase, with the active involvement of an adviser from the Finnish Embassy. The strategy cross-cuts all project activities, including the recruitment of project staff, and the project budget was later revised to allocate resources to its implementation. It notes that, beyond project-level interventions, the project cannot seek social change in project communities without also addressing [GESI] issues internally [within the project]. It reports on the actual situation in the project area and identifies the practices, values and norms used to justify and rationalise discriminatory practices. While the degree and forms of discrimination vary among communities, the practices are always based on a mixture of faith, fatalism and fear that is maintained by traditional beliefs. Eradicating this, the strategy concludes, will require a process of social mobilisation, sensitisation and confidence-building that will take a good deal of time and, in some cases, generations.

"In line with the rights-based approach, project personnel see the process of integrating cross-cutting issues and applying the GESI strategy as being just as important as the end results foreseen in the original project document. The project has therefore developed a comprehensive and participatory monitoring system that functions at all levels, one aim being to provide disaggregated data on the extent to which women and marginalised ethnic groups are involved in project activities and decisions. It also seeks to identify barriers to resources, motivations and experiences, and ways to reduce or remove these barriers. The participatory process used by the project has already sensitised stakeholders widely. Central and local government officials reportedly see the project as a good example of how to put into effect the principle of inclusion that is expressed in government policies. With Ministry of Local Development support, a large World Bank programme was recently prepared basing its implementation design on RVWRMP practices."

Source: Appendix 1 (33); quotation from Caldecott, Hawkes, Bajracharya & Lounela (2012: 64–66).

The CCTs are a priority of many aid policies. For example, the European Consensus on Development (EU, 2005) lists such priorities as the promotion of human rights, gender equality, democracy, good governance, children's rights and indigenous peoples, conflict prevention, environmental sustainability and combating HIV/AIDS. Meanwhile, Finland has been experimenting with the CCTs since 1993, starting with equality, democracy, environment and human rights, and every few years adding more until the CCTs included good governance, democratic accountability, rule of law, human rights, gender equality, HIV/AIDS, and the functioning of political parties and parliaments, with environmentally sustainable development as a parallel objective. Every part of this list can be unfolded into a mass of detail, including definitions, practices, indicators and lessons learned, along with the understanding that relative priorities evolve over time. The approach is hard to comply with in practice, and equally hard to score in evaluations (Box 2.21).

Box 2.21 Mainstreaming the cross-cutting themes of development cooperation

"... mainstreaming fails because the officials responsible are stressed and short of time, there is no time to read things carefully, the CCTs are not understood and there are no tools that would help people to learn about them, nor knowledge about when and where to ask for advice. There is, moreover, no accountable institutional anchorage for ensuring, and no systematic and obligatory practical arrangements for assuring, that the CCTs are embedded within Finnish activities, especially at the earliest stages of identification and design when they could make a strategic difference to the unfolding project, or at inception when their absence might be noticed or corrected. As a result, by the implementation stage they will have already been forgotten, and no baseline will have been established or measurements taken that could remind anyone of their existence or show that any progress has been made. In the absence of automatic procedures to ensure compliance, or comprehensive training and easy-to-use tools like guidelines and help systems, the influence of the CCTs on government aid activities (as opposed to NGO ones) can only be expressed through the personal interest and enthusiasm of the individual desk officer, embassy official or adviser. This is occasionally enough, of course, but as a system-wide response to a system-wide set of responsibilities, it is clearly inadequate" (Caldecott *et al.*, 2010: 77).

Sources: Appendix 1 (10–31).

In principle, complex, systemic issues that affect, underlie or could undermine everything – such as biodiversity loss and climate change – should be addressed in a mainstreamed and integrated way. Thus they should be taken into account in all decisions and addressed by building capacity to appreciate and respond to them in all contexts, rather than being targeted through stand-alone interventions. This makes superficial sense, but mainstreaming is hard work, as shown by the CCT experience, and the risk is that a lack of focus and technical backing will lead to nothing much happening in practice. In addition, there is the suspicion that calls for mainstreaming and integrated approaches, without specific measures and targets, are sometimes used to divert attention from systemic issues and placate those with a special interest in them. The converse challenge is that even if resources are explicitly targeted on systemic issues, they may be assigned to specialists and special programmes, which might then become isolated from the mainstream and 'ghettoised'. Hence there is a tension between these reactions to systemic risk, for which no easy remedy is available. The answer may lie in a mixed strategy, in which all aid professionals and their partners in development are:

- *alerted proactively* (i.e. through policy statements and training) to the nature and significance of all forms of systemic risk;
- *provided with access to guidelines* and digital help systems to support better understanding of the various risks and options for mitigating them;
- *required to describe early in the project cycle* the specific ways in which an intervention is vulnerable to systemic risks and the measures by which they are to be addressed;
- *informed by evaluations* throughout the project cycle, with explicit reference to systemic risks under the criteria of relevance, sustainability, coherence, and connectedness; and
- *informed by periodic analyses* of the strategic context of the programmes within which they work, including explicit reference to systemic risk factors.

In addition, there are lessons to be learned from holistic efforts to shift entire economic systems towards greater sustainability. These include support to businesses on environmental and social risk management, the presence of national cleaner production centres, and the availability of green credit mechanisms, which can be used as individual interventions but have far greater potency when deployed together. And even these are just a start, since there will also be a need for supportive fiscal incentives, educational networking at all levels, policies and leadership visions for low-carbon, biodiversity-friendly sustainable development pathways, and the unleashing of consumer preferences for demonstrable environmental friendliness across whole production and supply chains. In other words, beyond the level of aid interventions, there is a universe of ideas and practices which a reinvented global economy will increasingly have to inhabit. But first, within the bounds of this book, there is a need to improve the evaluation and design of the aid investments that safeguard resources while testing the new ideas that will be needed in the immediate future.

Design quality

The 2010 synthesis evaluation of the Finnish aid portfolio examined activity design, and found it to be consistently the lowest-scoring of the evaluation criteria, leading (it was thought) to other consistent weaknesses in impact, effectiveness and sustainability. Many factors were considered in scoring activity design (Box 2.22), on the grounds that "none of these features is optional if an activity is realistically, adaptively and efficiently to deliver useful results and sustainable impacts" (Caldecott *et al.*, 2010: 23).

Box 2.22 Factors considered in scoring activity design

"The bottom-most score for activity design is significant, since it is influenced by the presence or absence of such key factors as: clear objectives and selection criteria; clear guidelines, adequate systems, and meaningful, measurable and measured indicators for monitoring and evaluating project achievements in terms of poverty reduction; adequate risk analysis and arrangements for risk management; adequate financial and economic analyses; realistic expectations and relevant assumptions; understanding of relevant legislation and administrative mechanisms; responsiveness to conflicts of interest among key stakeholders; well-developed logframes that include realistic assumptions and risk analysis, and that are useful and used effectively in activity management; clear and flexible exit strategies; alignment to and management of stakeholder expectations; documented analysis of obstacles and discussion about alternative ways to deal with problems; a participatory approach that actively involves key stakeholders; an appropriate balance in focus between technical inputs, activities and achievements, and impacts and capacity-building collaboration with implementation partners; adequate arrangements for knowledge management to support the learning of lessons and dissemination of findings and conclusions; clear links between activities and intended results, based on adequate information on all the socioeconomic and ecological systems on which they depend or which are likely to be affected; adequate scope for involvement and reward of local suppliers of goods and services; and proactive use of information to ensure that the activity targets the poor as much as possible" (Caldecott *et al.*, 2010: 22–23).

Sources: Appendix 1 (10–31).

In the 2014–2015 evaluations it was decided that the activity design criterion was too broad, so most of the individual factors that were thought likely to influence performance were distributed for attention under other criteria, or else addressed in accompanying text. The approach continued to be represented, however, in the form of the following design quality criteria:

- *evidence and reasoning*, in which a high score is given if a design provides a convincing analysis of the context, problems, needs, and risks upon which it is founded, and presents sufficient evidence that its approach can deliver useful results and sustainable impacts (see Boxes 2.23 and 2.24);
- *clarity of explanation*, in which a high score is given if a design conveys a complete sense of how and why the project should be implemented (see Boxes 2.25 and 2.26); and
- *participation in design*, in which a high score is given if it is clear that all key stakeholder groups have been consulted in the design process, and have influenced it enough that their support can plausibly be expected (see Boxes 2.27 and 2.28).

Considering and scoring these topics can help to remind aid agencies that transparency and accountability require their arguments to be clearly marshalled and communicated, reports to be straightforward, and consent to be confirmed and ratified by on-going participation among those affected by their investments. These design quality criteria were not used in forming judgements on performance according to the core criteria, partly because the clients considered them subjective and unfair, so they were used as adjunct criteria instead. Unravelling the precise linkage between design and performance in aid portfolios is likely to reward further study as the science of evaluation evolves.

Box 2.23 Design quality (evidence and reasoning) – a high score in Vietnam

Environmental and social (E&S) risk management. "The programme is well-founded upon baseline studies that established the principles for the intervention. For example, surveys showed that banks were motivated to protect their reputation and avoid devaluation of collateral because of environmental non-compliance by clients, and that the most important constraint hindering banks' consideration of E&S risks was the absence of specific guidelines and other guidance material for the financial sector to help banks identify and manage related E&S risks, with other important constraints being the lack of qualified internal staff with relevant E&S expertise and experience, and the absence of qualified and affordable environmental consultants, all of which are deliberately targeted by the programme" (Caldecott *et al.*, 2014: 292–293, project UR–00593).

Source: Appendix 1 (46).

Box 2.24 Design quality (evidence and reasoning) – a low score in Tajikistan

Khujand Water Supply Project Phase II. This investment was assessed by the donor as 25 per cent relevant to mitigation and 25 per cent relevant to adaptation, but "the credit proposal does not in any way refer to climate change: not as a potential risk due to experienced and/or increasing weather extremes, nor as an issue to which the project would contribute as a solution – be it through mitigation or adaptation" (Caldecott *et al.*, 2014: 382, project UR–00174).

Source: Appendix 1 (43).

Box 2.25 Design quality (clarity of explanation) – a high score in Bangladesh

Agro-Forestry Improvement Partnership. "The credit proposal is clear, explaining the context and objectives in a solid and well structured manner" (Caldecott *et al.*, 2014: 324, project 7F–03804).

Source: Appendix 1 (42).

Box 2.26 Design quality (clarity of explanation) – a low score in Mongolia

Coping with Desertification Project (CODEP). The logical framework for the main phase of CODEP, which began in May 2008, described the outcomes, outputs, actions and indicators for the four project components, but there was no clear rationale for the components or their sub-components, and no explanation of how they related to one another, why they should all be addressed in one project, or why they were relevant to climate change (Caldecott *et al.*, 2014: 149, project 7F–05405).

Source: Appendix 1 (40).

Box 2.27 Design quality (participation in design) – a high score in Nepal

Hill Maize Research Project. "In a 15-year engagement with scores of districts and tens of thousands of farming households, and with arrangements in place to promote beneficiary participation and to obtain feedback from beneficiaries and link this both to policy and research direction, the extent of participation in project design must be considered to be very good" (Caldecott *et al.*, 2014: 243, project 7F–03093).

Source: Appendix 1 (42).

Box 2.28 Design quality (participation in design) – a low score in Vietnam

Commodities Certification Tropical Timber, Phase II. "The result areas are described as including a group of small-holder producers in the project activities but the group was yet to be identified in the credit proposal. Industrial plantations and communities are also mentioned as project partners, but their role in project implementation is not clear, and nor is it clear if they have participated in project design. It is worth noting that according to the credit proposal, the project mainly targets State Forest Enterprises, thus local communities are not directly involved in project implementation either" (Caldecott *et al.*, 2014: 297, project UR–00015).

Source: Appendix 1 (46).

Performance of aid institutions

Aid institutions have distinctive cultures, as a result of influence by their founders, leaders, or key experts at formative moments, the authors of guidelines that they use, or accidents and random events that occurred in the past. The British aid agency DFID, for example, has a style different from that of any other, in routinely emphasising the aspiration of transformative outcomes, and the use of counterfactual analysis and theories of change to justify investments in the form of a 'business case' for each. The value of these distinctive requirements becomes clear when a project is encountered that is under-ambitious, potentially inefficient, and incoherent or illogical due to inadequate attention to these very factors. This is not to say that DFID is uniquely virtuous in all things, but only that all agencies do things in their own way, which may be strengths or weaknesses depending on circumstance and context.

Areas of weak performance can sometimes be traced back to capacity weaknesses within the institutions that oversee or finance aid investments. In the

Finnish programme, for example, consistently low scores for complementarity were attributed to weak collaboration arrangements among partners, while a similar problem with replicability was linked to poor knowledge management (Caldecott *et al.*, 2010). The latter is shorthand for a cluster of arrangements and capabilities that make it possible for the staff of an institution to access, use, and learn from all the mistakes, successes, and insights that have gone before. Many factors contribute to this, from archival management, referencing, and synthesis, to staff training, turnover, and institutional memory. High staff turnover rates were said to be constraining performance at MFA Finland's geographical departments in the early 2000s (CDI, 2002), and informants were reporting the same problems a decade later (Caldecott, van Sluijs *et al.*, 2012; Caldecott, Valjas *et al.*, 2012). In any case, knowledge management was known to be a problem at MFA Finland and was being reviewed at the time of the 2010 evaluation, so the latter looked only at the possible utility of digital help systems in relieving learning and referencing issues (Box 2.29).

Box 2.29 Digital help systems

"[Help systems are] all kinds of learning and referencing systems that include useful knowledge organised in the form of browse paths and decision-support menus and keys, with pop-up definitions of key terms and supported by a searchable database of key words, concepts, case studies, regulations and procedures. One or more help systems should be prepared to assist in the mainstreaming of critical enabling factors (i.e. 'cross-cutting themes') and environmentally sustainable development (i.e. harm prevention and benefit capture from climate change mitigation and adaptation, water and food security, environmental education, community empowerment and fate control, biodiversity conservation, and humanitarian assistance). All of these areas currently suffer from high transaction costs due to unshared vocabulary and knowledge among development actors, a lack of familiarity with concepts, options, procedures and best practices, and the difficulty that time spent finding this information competes with other demands placed on overstretched personnel. Digital help systems offer a way to relieve these constraints" (Caldecott *et al.*, 2010: 116).

Sources: Appendix 1 (10–31).

How do aid institutions and their partners learn from what they do, and adapt over time? This question is most clearly asked in relation to the core criterion of replicability, where it concerns "the learning of lessons, the understanding of processes and relationships" (Table 1.2) and the "dissemination of findings and conclusions" (Box 2.22), but it has much broader implications. Many factors are involved, and all should be paid attention to. These can all be called 'memory systems', since they all hold traces of past performance that can influence the future, and all can help to determine sustainability and impact:

- *institutional memory*, which exists among the people that an institution employs and deploys, and within the archive-search-retrieval systems that it uses to support its work;
- *network memory*, which exists within global archives on development experience, in the form of reference collections of project documents, situational analyses, review documents, synthesis evaluations, and databases (such as CRIS);
- *capacity memory*, which exists among partners in the form of the laws, policies, procedures, maps, strategies, baselines, systems, relationships, capabilities, attitudes, forums, and other results of 'capacity building' that have been left behind by past cooperation;
- *physical memory*, which exists in the roads, buildings, dams, equipment, and other things created or installed as a result of past cooperation, along with the knowledge that they generate during their operation, maintenance, disuse, and disposal (and the impacts linked to these processes);
- *observer memory*, which exists among individual stakeholders who have witnessed or been involved in previous aid investments, including consultants, people associated with international or local NGOs and community-based organisations (CBOs), and local people affected by them; and
- *ecological memory*, which exists in the lasting effects of development activities on forests, grasslands, soils, ground waters, rivers, coastal, and other ecosystems, and the lifeways of people and other organisms that depend on those ecosystems, whether these effects are recognised and understood by development practitioners or not.

Of these categories, the most widely accepted as important are institutional and capacity memory, the former because its operation and maintenance is a continuing and often unsolved challenge that is widely recognised as such, and the latter because capacity building is a stated aim of most investments. Network memory is an area that may need further work, since it would be helpful for all concerned if all documents generated by the aid community were to be digitised and filed somewhere and accessible to global researchers, a task that might be funded through a small levy on all aid investments. Meanwhile, physical memory is usually mentioned under different names, but there are perhaps weaknesses in learning from the use and maintenance of installed facilities and particularly from their fates after the project that created them is completed.

Observer memory is used if people and groups with relevant experience are recognised and consulted (as 'consultants' or 'stakeholders') when investments are being formulated, but in the broader sense it may only come to attention when people object to a development plan because of a legacy of distrust left by previous investments. This may overlap with ecological memory if the impacts on people are perceived by them to have been significantly negative and due to environmental damage, or to the loss of access to ecological resources. Examples include the lasting effect of the 'Mega-Rice' project in Indonesia in the 1990s (Box 2.30), which nearly two decades later was still affecting public attitudes both to REDD+ programming in Central Kalimantan where it occurred, and to plantation development in Papua (Box 2.31). The fact that people in Papua had learned from events long before in Borneo illustrates the new interconnectedness of peoples, another example being the spread of concerns over the impact of timber harvesting that were being expressed in advance of the logging frontier in Sarawak in the 1980s (Box 2.32). One general point is that ecosystems and people store information about development investments which can be accessed and used, increasingly be everyone, and these can affect later plans. Another is that even climate change can be seen as a 'memory' of past GHG emissions, which the present and immediately future generations experience and to which they must react.

Box 2.30 Ecological memories in Indonesia

"The disastrous Mega-Rice Project, perhaps the largest and most destructive agricultural conversion project in the world in recent times, demonstrated how rapidly areas of orangutan habitat can be destroyed. In a bid to boost the country's rice production, one million hectares of peat-swamp forest was partly cleared and drained during 1995–97 in preparation for conversion to rice fields. Most of this land is covered in highly acidic, deep peat and is useless for agriculture. The construction of a network of massive canals completely drained the peatland during the dry season, and even when it became apparent that rice wouldn't grow and the project was abandoned, the drainage of the proposed rice field areas also drained vast tracts of the surrounding forests. Dead wood and dry peat became a tinder-box, flaring into uncontrollable fires that raged for six months during 1997–98. Over 400,000 hectares of forest burnt (Page *et al.*, 2002) and virtually no forest remained for orangutans to seek refuge in. In any case, the canals, rivers and farmland largely prevented orangutans from moving into the remnant forests. If we estimate an approximate orangutan density for the area of 2 individuals per square kilometre, that equates to 8,000 individuals that perished in the fires. A wasteland is left where before there was diverse rainforest."

Source: Singleton *et al.* (2004: 170).

Box 2.31 Observer memories in Indonesia

Central Kalimantan. "The [REDD+] Task Force aims to create a strong foundation from which to identify priorities for REDD+ implementation in Central Kalimantan, in line with the five pillars of the Strategy. The focus is on trust building and developing local ownership so that progress on the ground can continue regardless of events at the central level. Because of the history of provincial-central dialogue over the financing of the rehabilitation of the ex-Mega Rice area, there is a legacy of distrust and the provincial government is seeking a legal instrument to authorise their participation in REDD+, to delineate their tasks, functions, responsibilities and rights, and to give them confidence in the relationship with central government" (Caldecott *et al.*, 2011: 27). "A particular issue in Central Kalimantan is the enormous extent of land degradation caused by the 'Mega Rice' project, a failed attempt to cultivate and settle a large area of deep peatland" (Caldecott *et al.*, 2013: 16).

Papua. "There is a general concern, widespread in the outer islands of Indonesia (i.e. the islands beyond the national 'core' islands of Java, Bali and Madura), about central government initiatives that result in sudden, large-scale land-use change. This arises from the historic cancellation of traditional rights to lands and forests in the 1960s, the allocation of those resources to central government control, and their uncompensated consumption by non-local interest groups ever since. An example of this process was the One Million Hectare Peatland Development (or Mega Rice) Project in Central Kalimantan, Indonesian Borneo. This involved the logging and draining of forested deep peatland and the movement of about 100,000 migrants into the area, where they were expected to grow rice, thus shoring up official targets for both assisted transmigration and rice production. Once logged and drained, the peatland became fire-prone as well as being unsuited to rice growing, so the initiative led to an ecological disaster that produced desertification, local extirpation of forest-dwelling species, the release of an estimated 150 million tonnes of carbon-based greenhouse gases (GHGs) in 1997 alone, and economic calamity for in-migrants and indigenous people alike. The Merauke Integrated Food and Energy Estate (MIFEE) initiative is of a scale that invites comparison with the Mega Rice Project, and with this as a precedent there is the expectation within Papua that most of the financial benefits flowing from the MIFEE will be captured by stakeholders outside the province, while the greatest environmental and social costs will be borne locally. There is also the probability of massive GHG release due to planned and unplanned forest conversion as the MIFEE is developed. The area is drought-prone and vulnerable to fire during El Niño events, and has burned extensively during the current severe El Niño" (Indrawan *et al.*, in press).

Sources: Appendix 1 (32, 36).

Box 2.32 Observer memories in Sarawak

"Perceived conflicts of interest between shifting cultivators and loggers flourished in the Baram from the mid-1980s onwards. This was partly because the longhouses of the upper Baram tended to be surrounded by intact rain forest, while those in the middle Baram had recently had their first experience of widespread logging. Although the area was large and sparsely populated, motorized river transport and government-subsidized air travel had greatly improved communications. This meant that the inhabitants of the upper Baram were able to compare their own lifestyle with that of their downstream neighbours who had already experienced logging. Because of this contact, longhouse-dwelling people in the upper Baram began to voice their concern about a number of issues.

"There was a widespread belief that logging concessions were being issued for the lands around longhouses without their inhabitants being consulted. Moreover, the holders of these concessions were seen as becoming wealthy, while Orang Ulu groups or individuals were not participating fully in these benefits. Environmental damage was also anticipated as a result of logging, which was expected to affect income, lifestyle and diet, and which would damage ritually important sites. It was also thought that monetary compensation for this damage would be inadequate or, if provided in the form of labour fees, inadequately distributed. Lastly, there was a fear of social and religious disruption caused by an influx of outsiders and exposure of young people to outside influences.

"A different kind of concern was expressed by the Baram Penans, who in places remained wholly forest-dwelling and without fixed places of residence. Many perceived logging as a direct assault on their way of life and on their idea of environment and their place in it. As Brosius (1993: 142) observed: 'By transforming the landscape, logging destroys those things that are iconic of Penan society. When bulldozers clear roads through the forest, they scour away the surface of the land and obliterate recognizable features. ... The cultural density of that landscape – all the sites with biographical, social, and historical significance – is hidden, producing a sort of collective amnesia'.

'By the mid-1980s, the scene was set in the upper Baram for a strategic dispute between two main groups with very different attitudes and aims. In summary, many of the Orang Ulu wanted some areas to be protected from logging, and also greater participation in the financial benefits of logging. They were joined for most purposes by some Penans, who wanted a secure area in which to live in a traditional way. The State Government and the timber industry, meanwhile, had an interest in gazetting and using large areas as production forest. Both sides had vociferous allies within the Baram, within Sarawak, elsewhere in Malaysia and in the outside world, and their debate became linked to domestic politics at the State and

Federal levels. Global publicity about logging in Sarawak was seen in Malaysia as a threat to the country's tourism industry, while calls for a boycott of Malaysian timber and a hostile resolution of the European Parliament (EC, 1988) were seen as threatening timber exports. As a result, logging became a very heated issue in Sarawak during the middle and later 1980s."

Source: Caldecott (1996: 17–20).

Improving internal performance by aid institutions is a key part of aid investment, but lies in the background relative to the evaluation of any particular theme or project in aid delivery, and even more so in relation to broader issues of memory and adaptation. It is often addressed through management consultancies to look at the institutions themselves, though how those studies learn from events in the field is itself a knowledge management challenge. In any case, some internal investments are encountered within aid portfolios that seek to address systemic weaknesses which are hard to classify otherwise. These include efforts to promote thematic cooperation among agencies and other stakeholders in the donor and investor community, for example on renewable energy and energy efficiency (Box 2.33), on tax collecting and fiscal incentives (Box 2.34), and on policing against corruption and money laundering (Box 2.35).

Box 2.33 REPIC – building coherence on climate change

Three Swiss government agencies (SECO, SDC and SFOE) collaborate in the Renewable Energy and Energy Efficiency Promotion in International Cooperation (REPIC) platform to develop relevant partnerships with private enterprise and Swiss civil society. A series of contributions to REPIC was scored 5 for mitigation effectiveness because of anticipated leverage effects.

Source: Appendix 1 (37); Caldecott *et al.* (2014: 10, projects 7F–01587 and UR–00123).

Box 2.34 TADAT – missing coherence on climate change

A Tax Administration Diagnostic Assessment Tool (TADAT) was developed by the International Monetary Fund (IMF) to establish a mechanism for identifying weaknesses in country fiscal systems, so that they can be reformed and lead to higher tax revenues for partner governments. In considering a contribution, DFID noted the lack of a TADAT criterion for assessing, and help systems for promoting, fiscal incentives for climate change responses, but the IMF preferred to leave such environmental issues to the World Bank and UN systems. Without such a directional component, TADAT was assessed as essentially neutral with respect to climate change and environmental degradation, and therefore amounted to a lost opportunity due to a lack of coherence among international development institutions.

Sources: Appendix 1 (49); Caldecott (2014a).

Box 2.35 EIAAC – enhancing international action against corruption

A DFID investment aimed to strengthen the capacity of the UK's Proceeds of Corruption Unit of the Metropolitan Police Service, the Crown Prosecution Service Asset Recovery Unit, the Overseas Anti-Corruption Unit of the City of London Police, and the Intelligence Cell on International Corruption in the Serious Organised Crime Agency, during the establishment phase of a new National Crime Agency. Money laundering and corruption are associated with many kinds of crime, including illegal trade in endangered species and wildlife products, illegal and/or undocumented trafficking in ozone depleting substances, dumping and illicit trade in hazardous wastes, illegal, unreported and unregulated fishing, illegal logging and the associated trade in stolen timber, illegal transfer of genetically modified organisms or potential alien invasive species, and criminal activity surrounding financing for REDD+ programmes and other carbon conservation investments. Direct benefits from the investment could not be quantified in advance nor linked through specific chains of causality to climate change and environmental protection, but it was assessed as having numerous potential and/or indirect benefits in these areas. They could be further increased by raising awareness on environmental crime among participating institutions, through training by institutions such as Interpol's Environmental Crime Programme, the UN Interregional Crime and Justice Research Institute, and UNEP's World Conservation Monitoring Centre.

Source: Appendix 1 (49); Caldecott (2014b).

Intrinsic factors in evaluations

Aid investments are rationalised, conceived, paid for, and implemented by people, and aid evaluations are commissioned by human institutions and undertaken by individuals. The usual biases and weaknesses therefore apply, and can influence evaluations so they should at least be noted even if they cannot wholly be neutralised. A few such issues are raised in Box 2.36, drawn from my own experience, and may help others in their efforts to work as effective consultants.

Box 2.36 Practical lessons learned from evaluations

Cross-validation of desk studies. Where multiple investments in a portfolio are being evaluated by several people, it is a good idea for at least some of the investments to be examined independently by more than one member of the team, and the results compared. Closely similar findings and scores obtained by two or more people for the same project or programme help confirm that the individuals have a shared understanding of the methods, and that real phenomena are being documented. This was done for several of the evaluations in Appendix 1, with reassuring results in all cases.

Ground-truthing of desk studies. Evaluations based solely on written material and remote interviews are vulnerable to speculation that different findings might have been obtained through site visits. It is therefore a good idea for at least some aspects of an evaluation to be ground-truthed by visiting places where a programme has operated on the ground. Field visits are expensive, however, and contain their own sources of sampling error (e.g. who among the national stakeholders happens to be available at the time of the visit) and bias (e.g. cultural disposition for and against face-to-face outspokenness), so they should be carefully justified rather than being done for their own sake. They nevertheless often add a nuanced understanding of the context and diverse perspectives and motivations of the stakeholders involved, so almost all the studies in Appendix 1 involved visits to countries and field sites, and usually confirmed that the desk work was adequately informative and not misleading. This also provided some validation for findings that rested solely on desk work, and more generally offers reassurance that desk studies alone can provide a lot of useful information, and sometimes all that is truly needed for a large-scale evaluation of a diverse programme.

Consultants and consulting firms. Individuals differ in more than their technical qualifications, but the consulting world is based on the assumptions that the practitioners of specialities are interchangeable, can be packaged reliably by consulting firms, and the packages selected objectively by clients on the basis of price per expert-day and less tangible features such as the quality of the proposal and team, and the reputation of the offerer. The tendency is for consulting firms to pretend that it is they, rather than the consultants who do the work, that are intellectually (as well as legally) responsible for the outputs, thus building a corporate reputation that they hope will bias the award of contracts in their favour. Meanwhile, there are few incentives for either clients or firms to recognise the performance of individual consultants, and the authorship of documents is anyway often confused by a multiplicity of co-authors. Even if they are not prevented from doing so by the rules, this makes it hard for clients to choose individual consultants to do jobs for which they are particularly well suited, or avoid those who under-perform, and this constraint inevitably degrades the overall performance of the system. Since this involves a three-way contest between individuals, firms and clients, all with different interests and constraints, it is hard to think of a way to resolve the problem fully. For an individual consultant, the solution can only be to build and protect their personal reputation for high performance as best they can.

Challenges for clients. For donor agencies, the need for their investments to be independently evaluated from time to time is a challenge. Each evaluation is managerially demanding and technically complex, and the officials responsible may be under-resourced for the tasks involved. There is also potential tension between the need for an evaluation to be independent of the institutions involved, and the wish of the leaders of those institutions not to be embarrassed by its findings. From the clients' point of view, some of these challenges can be relieved by soliciting bids for evaluation contracts from third parties, which are often consulting firms outside the civil service which compete on quality and price with one another. This out-sources much of the effort and offers some independence and value for money, but for the commissioning agency there is still the risk of embarrassment if the evaluators find something wrong.

Controlling evaluations. To avoid potential embarrassment, clients might be expected to wish to control the evaluations of their investments, and it is often possible to interpret arrangements as being consistent with this. Control measures can include writing the ToR to steer the evaluators towards some issues and away from others, imposing rules that restrict access to information, and requiring that the evaluator's report is subject to review and not paid for until it is of acceptable 'quality'. Another measure that has the effect of controlling an evaluation is for the client to appoint one or several committees of 'experts' from other institutions to supervise it. The involvement of such committees can mean that the evaluators never know to whom they are actually reporting. Each advisory committee and its members may be beholden to the client because of the fees paid or the hope of future benefit, and this could compromise the integrity of advice that the committee may give the client against accepting certain findings, which may have to be changed before the evaluators are paid. This can exert powerful leverage over the evaluation, which might result in pressure on the consulting firm by the client, and on individual consultants by the consulting firm, ultimately to the detriment of truth and clarity.

Safeguards for evaluators. Two critical safeguards may be considered by those faced with signing an evaluation contract. Although the unequal power relationship between clients and evaluators can make it hard to comply with these, contractors should be aware that failure to use them can lead to professional frustration, moral hazard, and other difficulties such as compromised independence (up to and including micromanagement by the client) and unjustified criticism (up to an including verbal abuse by committees whose members are defensive of their own interests). The first safeguard is that the identity and role of the client must be clear – there must be one entity represented by named individuals which is responsible for commissioning the work and paying for it, on the clear understanding that no payment will be withheld if the ToR have been complied with. The second is that no entity other than the client itself should have anything other than a strictly advisory role, and that role must be in relation to the commissioner of the evaluation, not the evaluators. These safeguards may not make much difference if the client has already hired a committee to write the ToR. In such cases the ToR may be burdened with diverse questions and issues raised according to the interests of numerous individuals or institutions, creating a 'fog' of evaluation duties that will make it hard to do the work or obtain clarity in the findings. In these cases, the evaluator should make use of a clause that is usually contained in tender documents, entitled something like 'Comments on the ToR', to make it clear in advance that they will reject any unreasonable, incoherent or redundant tasks that distract from the focus of the evaluation, and any attempt to introduce advisers ('oversight committees', 'reference groups', etc.) with unspecified authority over the task.

Ideal tenders. A tender document should contain *only*: (a) a description of the object of the evaluation; (b) a list of core criteria against which it is to be judged (preferably the eight described here); (c) a limited number of special issues that the evaluators are asked to consider in the course of the research; and (d) an assurance that the evaluators will have free access to any and all information that they consider necessary to the task (subject to reasonable confidentiality rules). Any deviation from this content should be cause for careful thought, negotiation, and if satisfaction cannot be obtained, rejection of the contract.

Source: Author.

Using criteria to assess and compare investments

Cases illustrating the use of core evaluation criteria, and partner satisfaction and design quality criteria, reveal judgements about on-going or recently completed projects or programmes. Hence they are appropriate to the monitoring, mid-term, and post-completion stages of the project cycle, when investments are to be understood and corrected if necessary, or learned from and the loose ends tidied up, as appropriate. The same techniques can be used in advance of implementation, when an investment is subjected to appraisal and, if necessary, re-design. This is because the design of an investment should contain sufficient detail for its performance to be anticipated, so judgements can be made against the various evaluation criteria. The main role of the design quality criteria is to detect whether enough information is assembled and properly presented in the proposal, and whether the investment is likely to enjoy enough local support to be effective. Partner satisfaction is an emergent property of implementation, so it can only be anticipated imperfectly and mainly from the 'clarity of explanation' and 'participation in design' dimensions of design quality, where weaknesses might be expected to cause confusion or resentment later on. Apart from design quality, the key action in an appraisal is to check compliance with the following requirements:

- that the intervention is *relevant* to policy priorities and *coherent* with the aims and activities of other actors, while its *connectedness* to external factors and influences is factored into its design;
- that its performance is likely to be tolerably *efficient*, and *effective* in relation to its specified goals; and
- that the implicit claims that it is likely to have a significant degree of *impact*, *replicability* and *sustainability* are valid and supported.

Table 2.1 provides the summary findings from six evaluations, which are not named since their identities are irrelevant here. Each column represents an individual investment at the time of its examination (all of them being at the same stage of the project cycle), and the score in each cell of the table is supported 'behind the scenes' by a short essay in an evaluation report, in which the score is justified using evidence and reasoning (as in Tables 1.3 and 1.4). An exercise of this sort draws attention to particular weaknesses in the performance or design of the investments within a programme, which can then be corrected through re-design, management review, additional resource targeting, or some other response. This usage is at the heart of aid evaluation, and particularly the approach advocated here, which values the examination and understanding of aid investments as whole, complex systems (i.e. projects, institutions, peoples, and local systems affected by the investment, all behaving in various ways) interacting with other whole, complex systems (i.e. the national and sub-national context).

Then again, if all six columns in Table 2.1 are compared, it becomes clear that some investments are performing or are likely to perform better than others – in this case, programmes A and D look weak while the others look strong. This can be useful intelligence, if funds are short and only the best investments are to be supported or avoid cancellation, or if resources for re-design are limited and must be deployed to where they will do most good. The idea that investments tend to

Table 2.1 Performance scores for six aid investments

Criterion	*Performance of each programme by core criterion (score 7 best to 1 worst)*						
	A	*B*	*C*	*D*	*E*	*F*	*Mean*
Relevance	2	6	5	5	5	6	4.8
Efficiency	3	5	4	4	4	5	4.2
Effectiveness	3	5	6	3	4	4	4.2
Impact	3	6	5	4	5	5	4.7
Sustainability	2	5	5	4	5	5	4.3
Connectedness	2	4	3	2	4	5	3.3
Coherence	4	6	4	4	6	6	5.0
Replicability	2	6	6	3	6	6	4.8
	Estimated quality of each programme by design quality criterion (score 7 best to 1 worst)						
Evidence and reasoning	2	6	6	3	5	6	4.7
Clarity of explanation	1	6	6	3	5	5	4.3
Participation in design	2	6	–	4	5	5	4.4
Mean score	2.4	5.6	5.0	3.6	4.9	5.3	
Overall rating	Weak to very weak	Strong to very strong	Strong	Moderate to weak	Strong	Strong	

Source: Author.

score more highly according to some criteria than others is reinforced by Table 2.2, and such a spread in performance among criteria can usefully draw attention to systemic strengths and weaknesses in whole portfolios of investment. Consistent strengths might be highlighted in communications to parliament or the public, or used as a basis for lesson-learning and replication, while consistent weaknesses could attract investment in measures to correct them, or at least prompt questions such as "*Why do my projects have such low scores in connectedness, efficiency and sustainability, and what can I do about it?*" Where millions of dollars in public investment are at stake in a programme that is (or should be) of great public interest, such findings can (or should) exert immense strategic influence.

It is also possible to use consistent criteria and scoring techniques to build up a picture of how an investment's performance changes over time. This could be done through a succession of evaluations of one programme, and an example of this is given in Chapter 6, in which an EU-supported regional biodiversity centre was assessed five times in 2002–2008 as it evolved. It can also be done at the portfolio level, using constant methods applied to investments disaggregated by start date relative to some major change in context. This was done for the Swiss climate change portfolio, by looking at projects beginning in 2000–2006 and 2007–2012. The 2007 watershed between these periods was defined by publication of a particularly influential study of the economic implications of

Table 2.2 Ranking of evaluation criteria across 22 components of the Finnish aid programme

Criterion	*Mean score*	*Factors that might contribute to rank order*
Relevance	5.6	Relevance can be assessed once, relatively easily, at the beginning of the project. Aid officials may pay special attention to highly visible problems and needs, and wish to comply with stated policies.
Coherence	5.2	Anyone formulating an activity would be expected to pay close attention to the donor's and partner's priorities.
Partner satisfaction	4.8	Aid officials are likely to be sensitive to the opinions of their hosts, and to desire positive feedback.
Compatibility	4.7	Anyone formulating an activity would be expected to pay close attention to the donor's and partner's priorities.
Finnish added value	4.6	This was a Finnish government priority at the time (from 2007) and was being actively sought in most of the studies reviewed.
Coordination	4.3	Partners must collaborate in identifying, formulating, and managing their joint activities, which demands special and sustained effort.
Impact	4.3	Mediocre scores may result from weak activity design.
Effectiveness	4.1	Mediocre scores may result from weak activity design.
Sustainability	4.0	Mediocre scores may result from weak activity design.
Replicability	3.9	Requires the learning of lessons, the understanding of processes and relationships, and the possibility that future activities can be modelled on improved versions of past ones, all of them challenges for the best knowledge management systems.
Complementarity	3.8	Partners must collaborate in identifying, formulating, and managing their joint activities, which demands special and sustained effort.
Efficiency	3.6	Efficiency is hard to measure or achieve in any complex enterprise that has only partly quantifiable inputs and outputs, and intangible influences on context and performance.
Connectedness	3.6	Not all factors that exert influence on a project, but over which it has little direct control, are routinely considered.
Activity design	3.3	A demanding formulation requirement, in which the activity must respond to its local and broader context, the threats and needs operating or anticipated in those contexts, and the arrangements for mitigating or adapting to risks, monitoring and responding to events, and steering the activity towards the delivery of results and impacts in a reasonably efficient manner.

Sources: Appendix 1 (10–31); Caldecott *et al.* (2010: 60, 66–68).

climate change (Stern, 2007), the release of the Fourth Assessment Report of the Intergovernmental Panel on Climate Change (IPCC, 2007), the Bali Conference of the Parties to the UNFCCC at which the Forest Carbon Partnership Facility (FCPF) was launched (see Chapter 5), and 2007 was also the year after which, according to witnesses, more attention than before was paid to climate change by the Swiss government's aid institutions. Findings are summarised in Table 2.3, which shows a distinct upward shift in effectiveness between the two periods, with the share of the budget covered by strongly effective investments (scores 5–7) increasing from 39 per cent to 55 per cent for mitigation, and from 23 per cent to 66 per cent for adaptation. This was attributed by Caldecott *et al.* (2014: 41) to a more than doubling of the share of the budget covered by high scores (5–7) for design quality, especially according to the 'evidence and reasoning' criterion. Thus the aggregate performance scores of whole aid portfolios can reliably detect systematic improvements and lead to explanations for them, in ways that can be used to build appreciation of the investments among politicians and the public.

Table 2.3 Trends in effectiveness within the Swiss climate change portfolio

Score	*Effectiveness*	*Mitigation (percentage of total budget among 235 projects)*		*Adaptation (percentage of total budget among 242 projects)*	
		2000–2006	*2007–2012*	*2000–2006*	*2007–2012*
7	Extremely strong	3	1	0	0
6	Very strong	25	24	9	34
5	Strong	11	30	14	32
4	Moderate	52	40	69	24
3	Weak	9	3	7	9
2	Very weak	0	1	1	0
1	None	0	1	0	0

Sources: Appendix 1 (37–46); Caldecott *et al.* (2014: 40).

The next few chapters

The rest of this book treats recent aid activities that include attention to sustainability as 'baby steps' that had to be taken before we became able to make real progress against the major challenges that face us. Chapter 3 briefly discusses the big picture of what aid is for and how to tell if it is doing any good, before focusing on the perception of a global crisis of sustainability that has emerged since the early 1970s and the operation of key environmental treaties since the early 1990s. Each treaty was designed to address a particular problem, and those on climate change (UNFCCC), biodiversity (the Convention on Biological Diversity, CBD), and desertification (UN Convention to Combat

Desertification in Countries Experiencing Serious Drought and/or Desertification, Particularly in Africa, UNCCD) are of special interest here. I summarise the nature of the challenge in each case, and explain how the treaties allow for investments in climate, desertification, and biodiversity to be tracked, using the Rio Climate Markers and the various actions listed within the CBD and UNCCD. Chapter 3 ends with a note of caution that solutions to strategic problems have not yet been found, and that progress is becoming harder to make in some ways and places.

Afterwards, Chapter 4 on mitigation technology, Chapter 5 on mitigation ecology, and Chapter 6 on adaptation, summarise the main kinds of investment that were found in the evaluations listed in Appendix 1, and provide examples of their performance scores. Where scores are given, they take the form in which 7 represents the highest-possible (i.e. extremely strong) performance, 4 a moderate performance, and 1 the lowest-possible (i.e. extremely weak, if any) performance. Intermediate scores for performance are self-explanatory, from very strong (6) to very weak (2). This usage is based on the principle outlined earlier in this chapter, but with a seven-point rather than a four-point scale because the intermediate scores turned out to be rather useful in practice. After reading the explanations of what the various scores mean in Chapter 1, the highlights of findings in this chapter that gave rise to different scores, and the further examples in Chapters 4, 5 and 6, the reader should be left with no doubt at all about what is called for in the systemic evaluation of aid performance.

References

Brosius, J. P. (1993) Penan of Sarawak. Pages 142–143 in *State of the Peoples: a Global Human Rights Report on Societies in Danger*. Cultural Survival and Beacon Press (Boston, Massachusetts).

Caldecott, J. O. (1996) *Designing Conservation Projects*. Cambridge University Press (Cambridge).

Caldecott, J. O. (2014a) *Climate and Environment Assessment: Support for the Tax Administration Diagnostic Assessment Tool*. Evidence on Demand, HTSPE Ltd (Hemel Hempstead).

Caldecott, J. O. (2014b) *Climate and Environment Assessment: Tackling International Corruption in the UK through Anti-Money Laundering and Anti-Bribery Enforcement*. Evidence on Demand, HTSPE Ltd (Hemel Hempstead).

Caldecott, J. O., Le Thac Can, Muhtaman, D. R., Scutt, B. & Tan, E. E. (2002) *Final Report of the Mid-Term Review of the ASEAN Regional Centre for Biodiversity Conservation*. EU Delegation (Manila).

Caldecott, J. O., Becker, A. & Udomittipong, P. (2007) *Final Evaluation of the Coastal Habitats and Resources Management (CHARM) Project*. EU Delegation (Bangkok).

Caldecott, J. O., van Praet, C. & Cai, M. (2009) *Final Report of the Mid-Term Evaluation of the EU-China Biodiversity Programme*. EU Delegation (Beijing).

Caldecott, J. O., Halonen, M., Sørensen, S. E., Dugersuren, S., Tommila, P. & Pathan, A. (2010) *Evaluation of Sustainability in Poverty Reduction: Synthesis*. Evaluation report 2010: 4. Ministry for Foreign Affairs (Helsinki).

Caldecott, J. O., Indrawan, M., Rinne, P. & Halonen, M. (2011) *Indonesia–Norway REDD+ Partnership: First Evaluation of Deliverables*. Government of Norway's International Climate and Forest Initiative (Oslo).

Caldecott, J. O., Hawkes, M., Bajracharya, B. & Lounela, A. (2012) *Evaluation of the Country Programme between Finland and Nepal*. Evaluation report 2012: 2. Ministry for Foreign Affairs (Helsinki).

Caldecott, J. O., Hawkes, M., Bajracharya, B., van Sluijs, F., Aguilar, B., Valjas, A., Killian, B. & Lounela, A. (2012) *Evaluation of the Country Programmes between Finland and Nepal, Nicaragua and Tanzania: Policy Brief*. Special edition 2012: 1. Ministry for Foreign Affairs (Helsinki).

Caldecott, J. O., Valjas, A., Killian, B. & Lounela, A. (2012) *Evaluation of the Country Programme between Finland and Tanzania*. Evaluation report 2012: 3. Ministry for Foreign Affairs (Helsinki).

Caldecott, J. O., van Sluijs, F., Aguilar, B. & Lounela, A. (2012) *Evaluation of the Country Programme between Finland and Nicaragua*. Evaluation report 2012: 1. Ministry for Foreign Affairs (Helsinki).

Caldecott, J. O., Rizki, D., Rinne, P. & Halonen, M. (2013) *Indonesia–Norway REDD+ Partnership: Second Evaluation of Deliverables*. Government of Norway's International Climate and Forest Initiative (Oslo).

Caldecott, J. O., Dornau, R., Gollan, J., Halonen, M., Saalismaa, N., Schilli, A., Simonett, O., Stuhlberger, C. & Tommila, P. (2014) *Technical Report on Effectiveness of the Swiss International Cooperation in Climate Change Mitigation and Adaptation Interventions 2000–2012*. Mandate of the Swiss Agency for Development and Cooperation (SDC) and the State Secretariat for Economic Affairs (SECO) (Bern).

CDI (2002) *Evaluation of Bilateral Development Co-operation between Nicaragua and Finland*. Evaluation Report 2002: 4. Ministry for Foreign Affairs (Helsinki).

EC (1988) Resolution on the Catastrophic Environmental Impact of Large-Scale Deforestation in Sarawak (East Malaysia). *Official Journal of the European Communities* **31**(C235): 196–198.

EU (2005) *Joint Statement by the Council and the representatives of the Governments of the Member States meeting within the Council, the European Parliament and the Commission on European Union Development Policy: 'The European Consensus'*. General Secretariat Note 14820/05 (DEVGEN 229, RELEX 678, ACP 155). Council of the European Union (Brussels).

Indrawan, M., Caldecott, J. O. & Ermayanti, E. (in press) Mitigating tensions over land conversion in Papua, Indonesia. *Asia and the Pacific Policy Studies*.

IPCC (2007) *Climate Change 2007: Synthesis Report – Contribution of Working Groups I, II and III to the Fourth Assessment Report of the Intergovernmental Panel on Climate Change* (edited by R. K. Pachauri & A. Reisinger). Intergovernmental Panel on Climate Change (Geneva).

OECD/DAC (1991) *DAC Principles for Evaluation of Development Assistance*. www.oecd.org/dataoecd/31/12/2755284.pdf (consulted 12 Sep 2016).

OECD/DAC (2010) *Evaluating Development Co-operation: Summary of Key Norms and Standards*, second edition. www.oecd.org/dataoecd/12/56/41612905.pdf (consulted 12 Sep 2016).

Page, S. E., Siegert, F., Rieley, J. O., Boehm, H-D. V., Jaya, A. and Limin, S. (2002) The Amount of Carbon Released from Peat and Forest Fires in Indonesia during 1997. *Nature* **420**: 61–63.

Singleton, I., Wich, S., Husson, S., Stephens, S., Utami Atmoko, S., Leighton, M., Rosen, N., Traylor-Holzer, K., Lacy, R. and Byers, O. (2004) *Orangutan Population and Habitat Viability Assessment: Final Report*. IUCN/SSC Conservation Breeding Specialist Group (Apple Valley, Minnesota).
Stern, N. (2007) *Stern Review on the Economics of Climate Change*. HM Treasury (Cambridge University Press, Cambridge).
Uusihakala, J., Caldecott, J., Killian, B. & Njagi, J. W. (2015) *Finland's Development Cooperation With Kenya in 2007–2013*. Evaluation report 2015: 2. Ministry for Foreign Affairs (Helsinki).

Chapter 3

Keeping human development going

Purposes and problems

The purposes of aid

Poverty originates in various kinds of exclusion, exploitation, disenfranchisement, and dispossession, which are driven and perpetuated by cultural and political norms within societies. Most aid can be thought of as aiming to correct such distortions for the benefit of the poor, or more generally, the vulnerable. Hence its purpose can be defined as being to deliver "the greatest benefit for the greatest number of the most vulnerable people" (Caldecott *et al.*, 2010: 111). Such an aim need say nothing about the choice of what to target for maximum effect in any given case, from among such options as health, water, food, business, family planning, or climate change, but it should include the time dimension to encompass issues of long-term and inter-generational equity, and it should guide interventions towards areas that can be justified in terms of benefit delivery to the vulnerable.

Both quantitative and qualitative factors must be considered in evaluating progress towards poverty eradication. This is because qualitative factors such as governance, relationships between genders, castes, classes, and age groups, and the causes of various kinds of vulnerability, are so influential that the whole process is only partly quantifiable. Poverty has declined in many developing countries (Box 3.1), but a challenge for evaluating this lies in the attribution of effects to causes. There are always questions close to the surface, for example: does aid itself reduce poverty, or does it leverage policy or technology changes that allow greater wealth creation? Or, are aid effects eclipsed by other changes, such as the growth of mobile telephony in rural areas, or factors like regime change, the growth of a globalised economy, or enterprise and ownership reforms, which cause so much wealth creation that poverty naturally evaporates? Worse, as argued by authors such as Easterly (2006) or Moyo (2009), is aid a spoiling influence which, by creating dependency, discouraging enterprise, distorting economic systems, and promoting corruption, actually deters the creation and equitable distribution of wealth?

Box 3.1 Progress on poverty

"Between 1990 and 2015 income poverty in developing country regions fell by more than two-thirds. The number of extreme poor people worldwide fell from 1.9 billion to 836 million. The child mortality rate fell by more than half, and under-five deaths fell from 12.7 million to 6 million. More than 2.6 billion people gained access to an improved source of drinking water, and 2.1 billion gained access to improved sanitation facilities, even as the world's population rose from 5.3 billion to 7.3 billion."

Source: UNDP (2015: 4).

Given these uncertainties, it might perhaps make sense to focus only on removing the obvious constraints on wealth creation, such as excessive regulation, taxation, corruption, and trade barriers, so that people will become richer, after which they will be better able to solve any residual problems of sustainability and equity. This approach would shift the focus away from any kind of organised altruism and towards arrangements that promote the free movement of goods, capital, people, ideas, and technologies within and between countries. From this point of view, it is unnecessary to worry about the 'moving parts' of society (culture, institutional and individual motivations and behaviour, etc.), except as sources of friction that require money, law, and sometimes force to lubricate, and even less important to be concerned over residual matters like ecology, equity, and climate change. The problem, however, is that some issues that are dismissed as *residual* in this approach, to be set aside for later, optional treatment, are in fact *existential*, and must be addressed automatically, early, and completely, or the whole system will eventually fail. The next section summarises the development of our collective awareness of this fact in relation to environmental deterioration. It implies that economic systems require active direction if their ability to create wealth is not to undermine itself, the human social fabric, and, most importantly, the biosphere upon which everything depends.

So, with all this in mind, we come back to the purposes of aid. Several such are mentioned in this book, as well as the one about delivering *the greatest benefit for the greatest number of the most vulnerable people*. Thus, the purpose of aid oriented to climate change mitigation is described as being *to preserve as completely as possible and for as long as possible the environmental conditions that humans have experienced since the last Ice Age*. It is also stated that aid is bound together with the issues of mitigation, adaptation, desertification, and biodiversity through the idea of *keeping human development going*. And finally, in Chapter 9, I suggest that the role of aid lies in building a *functioning, inclusive, and ecologically lawful community among all people*.

In any case, the organised, politicised altruism of the aid process, the shadowy influences of private investment, the intellectual challenge of working out what is actually going on, why, where it is heading, and with what consequences, most of which are unknowns, all combine to make aid evaluation both fascinating and frustrating. Above all, the concept of sustainability has now become established, in particular with the idea of sustainable development being offered as an overarching but poorly defined goal, and also in the vocabulary of donor agencies that are interested in whether the effects they may or may not achieve using their aid budgets are capable of being sustained once any particular intervention is over. Thus, aid investments are occurring in what is increasingly recognised as a very complex and connected world, so the evaluation of aid will need to adapt accordingly. Moreover, with limited public interest in the complexities involved, and the risk of political changes and budget cuts, aid will also need to make the case for its own continuation based on much deeper, broader, and more thoughtful analyses than in the past, as well as a much clearer explanation of its goals, achievements, and relevance.

It is also worth contemplating how overseas aid relates to the other public investments of the societies that give it. The purpose of some portion of the UK's annual social protection budget of over £200 billion (more than the whole ODA budget of the world), for example, must be analogous at least to some of the purposes of overseas aid. Large items within it include various credits, benefits and allowances designed to relieve poverty, build capacity and enable people to find work. Many other items of government expenditure address the education, training, infrastructure, farming, health, nature conservation, and other agendas reminiscent of the targets of overseas aid, and use similar means of delivery, including the civil service, consulting firms, partnerships with NGOs, universities, etc.

After allowing for historical and political legacies, it may still be appropriate to ask whether the performance of all this internal public investment might be evaluated using the same tools as are proposed here for overseas aid, and whether there are lessons from aid investments that might be useful in a domestic context. For example, we know from cases in the Amazon, Costa Rica, Indonesia, and elsewhere, that circumstances can ripen until relatively small, highly targeted investments can induce transformational change, and that such investments show extremely high levels of performance (Chapter 8). We also know from various global, regional, and national cases that many years of consistent, adaptive, strategic effort can often eventually solve even the most intractable problems (Chapter 9). Those who direct and evaluate a government's investments in its own citizens can presumably be trained in how to detect ripening circumstances, in how to use them to induce transformations, and in how to justify a strategic approach to challenges, while individual investments and partnerships might well be evaluated using the criteria of performance and satisfaction described here.

Human development and environmental change

As the goals of overseas aid have been debated over the years, the term 'human development' has been used to focus attention on improving the lives that people actually lead, with income growth being seen as a means rather than an end, and with balanced attention to health, knowledge, participation, environmental and physical security, and equity. Its environmental dimension is rooted in events since the 1960s, which saw the early growth of an awareness that economic and population growth in a finite world could not be sustained for ever (e.g. Meadows *et al.*, 1972). Discoveries since then have broadly confirmed this conclusion, as numerous environmental constraints on development have been identified and explored in public debate, ranging from general resource depletion (e.g. Meadows *et al.*, 1992), destabilisation of the biosphere (e.g. Lovelock, 2006), implications of population growth (e.g. Hardin, 1993), and climate change (e.g. Houghton, 2004), to such more specific problems as the future of fisheries (e.g. Clover, 2004) and water supplies (e.g. Pearce, 2006). Leaving aside the effects of all this on poverty and malnutrition due to failing agriculture and employment, displacement of families and communities to locations ill-prepared to receive them, and physical damage due to civil strife and warfare aggravated by environmental change, there are also specific health and water issues (Boxes 3.2 and 3.3) that drive home the particular implications of environmental deterioration for the general public.

Box 3.2 Health and environmental degradation

Lung damage and increased susceptibility to respiratory diseases caused by air pollution are joined by other human health issues that arise from environmental degradation. These include: infectious diseases promoted by climate change, whether vector-borne (e.g. malaria) or water-borne (e.g. cholera); infectious zoonotic diseases promoted by forest degradation and hunting (e.g. Ebola); and infectious diseases promoted by global travel, whether vector-borne (e.g. Zika) or air-borne and zoonotic (e.g. bird flu). There are also local and trans-boundary effects of industrial pollutants, such as CFC-driven ozone depletion (e.g. skin cancers, actinic keratoses and premature aging of the skin, cataracts and other forms of eye damage such as macular degeneration, and suppression of immunity), the dumping of electronic and other hazardous wastes (acute and chronic disorders from exposure to heavy metals and other diverse toxins), and exposure to persistent organic pollutants including pesticides (causing disruption of the endocrine, reproductive, developmental and immune systems, nerve and brain damage, and induced cancers).

Sources: McMichael *et al.* (2003); Myers (2009); UNEP (2012a).

Box 3.3 Water and environmental degradation

Almost all the world's water is salty, and for us on land a regular supply of fresh, clean water is uniquely precious. Such a supply is the most important service that an ecosystem can offer, yet is often allowed to lapse through abuse, or it is diverted from those who really need it, or destroyed by over-use or pollution. Global consequences include a huge waste of human energy in an endless quest for water, and a huge burden of unnecessary illness. This global issue is made up of tens of thousands of local water crises, each one due to decisions that affect local ecosystems. Water and ecosystems are linked, from the boundaries of each catchment to the streams, rivers, lakes, floodplains, swamps, and estuaries created by water. All actions in a catchment have consequences for what happens down-stream, so logging, farming and grazing, applying fertilisers or pesticides, dumping garbage, releasing sewage, or spilling chemical wastes, all have impacts that are carried by water. Meanwhile, we pump water from the ground at rates far higher than its rate of replacement, causing wells to run dry in region after region. And the local ecosystems that determine local water supply also influence the occurrence of rain or drought, storm or famine. The changes we make to ecosystems are contributing to mass extinction, local water crises, and climate chaos that further stresses ecosys-tems, creating a dangerous cycle of effects that amplify causes, and causes that amplify effects.

Sources: Caldecott (2008); UNEP (2012b).

Set against all this, the last few decades have shown that human ingenuity can find ways to reduce demand for particular resources, to reduce the environmental impacts of particular industrial operations, to generate more value from particular inventions and transactions, and to distribute the benefits more inclusively among people (e.g. Monbiot, 2006). Meanwhile, it was also realised that weaker stakeholders in the globalising economy, whether individuals, families, peoples or countries, tend to be excluded from protection and advancement unless there are specific measures in place to promote their interests (e.g. Chambers, 1983). This applied especially to those who are politically marginal (e.g. Oxfam, 2007), often including the poor, jobless, rural communities, indigenous peoples, women, children, and non-human species, all of them relatively or entirely voiceless and vulnerable to the side effects of economic growth (MA, 2005a).

Over the years, there has been a steady growth in our collective understanding of environmental change at all levels (e.g. MA, 2005b): how it is induced, how it affects different groups, and how it can be mitigated or adapted to. People have continuously synthesised scientific and national reports into global reviews, such as the *Assessment Report* series published by the Intergovernmental Panel on Climate Change (IPCC) in 1990, 1992, 1995, 2001, 2007 and 2013–2014, the *Global Environmental Outlook* reports by UNEP (the fifth in 2012), the *Human Development Report* series by UNDP (annually since 1990), the *World Resources Report* series by the World Resources Institute (every year or two in 1986–2011, then every few months on selected themes), the *Living Planet Report* series by WWF (the latest in 2016, but reporting on species population trends since 1970), and others. International conferences on environment and development also periodically offered global-level policy and leadership perspectives: 1972 in Stockholm, 1992 in Rio de Janeiro, 2002 in Johannesburg, and 2012 again in Rio ('Rio+20'). By these means, an overview of global environmental circumstances has been maintained and renewed, allowing its main dimensions to be identified and to an extent addressed. These include the phenomena listed in Chapter 1, which taken together make the human condition unsustainable.

Treaties and challenges

There are a number of environmental treaties in force, and three are of special relevance to aid performance and climate change: the UNFCCC, the CBD, and UNCCD. All three were signed at the Rio conference in 1992, so together they are known as the 'Rio Treaties'. They help to organise our collective thoughts on, and responses to, the interconnected challenges of climate change mitigation and adaptation, biodiversity loss and conservation, and desertification.

Mitigating climate change

In ordinary English, 'mitigation' means *making something less severe*. When talking about climate change that is caused by people due to our emissions of greenhouse gases (GHGs), mitigation means reducing the amount of GHGs in the air, by any means that work. Its ultimate purpose is to preserve as completely as possible and for as long as possible the environmental conditions that humans have experienced since the last Ice Age, which have proved so amenable to the growth of farming, cities, and human populations. The underlying assumptions, both of which seem valid, are that any significant change in these conditions will be difficult for humanity, and that the main source of change is excessive net GHG emissions. Four main approaches are available to reduce these emissions:

- *directly and simply*, to make fewer emissions by substituting low-carbon technologies for high-emission ones;
- *directly but complexly*, to make fewer emissions by reducing the release through land-use change of carbon stored in living ecosystems, a process that can be folded into a whole dynamic of low-carbon development planning, or advanced at a landscape level through 'reducing emissions from deforestation and degradation' (REDD), and organic farming;

- *directly but innovatively*, to offset emissions that we continue to make or to reduce the effects of past emissions by using 'negative-emission technologies' (NETs), essentially to take GHGs out of exhausts or from the air, and make them safe by storing them out of contact with the biosphere; and
- *indirectly, complexly, and innovatively*, to encourage investment in low-carbon technologies, REDD, NETs, etc., by applying financial incentives, whether fiscal (driven by government interventions) or market-based (driven by consumer preferences), to reward and encourage 'good' behaviour, and to punish or deter 'bad' behaviour.

The first two approaches have been explored through enough aid investments that some good practices and lessons learned can be assembled, which is done in Chapter 4 for low-carbon technology substitution and Chapter 5 for REDD and related methods. The total amount of GHGs not released so far, however, as a result of investment in low-carbon technologies, REDD and organic farming, is very small compared with the sum of emissions from all sources – on the order of millions rather than billions of tonnes of carbon dioxide equivalent (tCO_2e) annually. These strategies are clearly part of the solution to climate change, and they also have important co-benefits, but they are not plausibly sufficient over the small number of decades (if that) before runaway climate change becomes inevitable and its consequences catastrophic. Hence the other two approaches, which have so far been explored in a very preliminary way, will need to be addressed with much greater urgency. The third (to do with NETs) is merely touched upon in Chapter 4, since the approach is still in its infancy despite its obvious strategic primacy, but a track-record that is relevant to the fourth (encouraging investment) is available and is treated briefly in Chapter 4. Otherwise, as the mitigation strategies mature, the other major priority is adapting as best we can to the changes that have already been induced and cannot now be reversed, which is addressed in Chapter 6.

Adapting to climate change

In ordinary English, 'adaptation' means *becoming adjusted to new conditions*. When talking about climate change, adaptation is mainly about 'resilience', which again in ordinary English means *being able to withstand or recover quickly from difficult conditions*. These difficult conditions are expected because a warming biosphere confronts us with an increasingly unstable world in which ever more numerous people are exposed, among other things, to:

- more severe storms, floods, droughts, heat-waves, and bouts of extreme cold;
- rising sea levels and associated coastal erosion and saline intrusion to fresh groundwaters; and
- progressive changes to, for example, marine and coastal ecosystem productivity as a result of ocean warming, currents and acidification of sea-water,

and long-term ice conditions on mountains that can desiccate or temporarily obstruct river systems, creating water shortages and acute flood risks.

Adaptation investments promote resilience to such local manifestations of global warming, and most of them are to do with the management of ecosystems such as forests and farmlands, and preserving or strengthening their ability to shield people from the effects of climate change while maintaining their productivity in terms of food, water, and other livelihood resources. Equally, adaptation must enable people to manage increasing intensity, scale, frequency, and unpredictability of natural disasters. Thus, an adaptation programme will typically involve building grass-roots and local government capacity for environmental management, sustainable agricultural systems, and disaster risk reduction, while also ensuring the adequate supply and distribution of safe, fresh water. Meeting these needs is an increasing focus of aid investments, and enough information on the performance of these is now available that examples can be given and lessons learned, which is done in Chapter 6.

It is important to be aware, however, that most adaptation investments focus on restoring or protecting local systems – a single catchment or irrigation system, perhaps, or flood-protecting a single town – but that adaptation and resilience ultimately depend on the health and behaviour of far larger systems, and in the longer term. Thus an adaptation investment may patch up a location, but more systemic work will also be needed, both on mitigation since otherwise climate change will overwhelm whatever is achieved locally, but also across whole landscapes and connected economic and ecological systems at the regional level – the Sahel, for example, or a whole river basin that may be as big as that of the Amazon or Nile. Moreover, adaptation must also involve the preservation and use of diversity, among crops, wild species and ecosystems, and among cultures, drawing on these resources to find ideas and techniques with which to promote sustainability under changing and unpredictable conditions. These larger-scale and longer-term investments can also qualify as adaptation strategies. Because all must take account of ecology and diversity, there is a very large overlap with two other themes of aid investment that are relevant to sustainability: combating desertification and conserving biodiversity, which are therefore also addressed in Chapter 6.

A final point is that climate change will inevitably cause the displacement of large numbers of people, from small islands and coastal zones mainly due to sea-level rise, salt intrusion and storm exposure, and within continental systems mainly due to drought and desertification, with flooding, disease, fire, and other hazards adding to the pressures. Population movements may be involuntary and sudden, or voluntary and slow, but will interact with policies, plans and laws at an increasing rate. The chaos in Europe resulting from the 2015–2016 mass arrival of desperate people displaced by conflict in south-west Asia draws attention to the need for careful contingency planning and the resourcing of adaptive measures. This would apply, for example, to such scenarios as the irreversible flooding of major cities, the creeping emigration of people who abandon life in water-stressed areas, and the failure of entire national economies among small-island developing states (SIDS) and deltaic countries. The starting point for planning will be different in each case, depending for instance on cultural, geographic, and historical factors (which will affect how easily a group of people can be accommodated in a new setting), and whether a policed national boundary limits the movement of displaced people. In all cases, while it is tempting to deny the problem and its causes, this response is not viable so coherent and acceptable solutions will need to be prepared for and budgeted. Debates over the share of resettlement expenditure that is properly classified as 'adaptation funding', and what are 'acceptable' solutions, will no doubt continue for decades.

Combating desertification

Drylands cover two-thirds of Africa and a third of North America, where in both cases almost three-quarters of the area used for agriculture is already degraded. Land degradation takes many forms, including nutrient depletion, salinisation, soil erosion, and ultimately desertification. Nutrient depletion occurs with continued cropping without sufficient nutrient input, and is made worse by the leaching effects of rainfall. Salinisation is a feature of over-irrigated farmlands, since mineral salts are first dissolved in soil water and then drawn to the surface by evaporation, where they are deposited in a toxic crust. Soil erosion is facilitated when vegetation cover is damaged by tree-felling or grazing, and can be driven by wind or water, often locally to create scarred and sculpted landscapes, but sometimes over very large areas as entire topsoils are blown or washed away. These processes all interact with each other, with land use, and with climate, and desertification often arises from a combination of over-grazing and poor farming practices to expose the soil, drought to weaken it, and wind to blow it away. Most of the drylands threatened by this are found near the six main desert areas of the world: the Sonoran, Atacama, Sahara, Gobi and Kalahari deserts, and most of Australia. Making nomadic people settle down in marginal drylands, or blocking their ancestral movements, can lead to desertification because they reduce people's ability to adjust their behaviour in the face of droughts. Problems of water scarcity, groundwater depletion, soil erosion, and salinisation can all be outcomes of such policy and institutional failures. The cumulative result has been that an area the size of China and India combined (about 12 million sq. km) has suffered moderate to extreme soil deterioration in the past 50 years, mostly in arid or semi-arid regions in developing countries. But combating desertification is hard to do, because of changes in rainfall patterns, land tenure that is no longer well adjusted to the environment, and the pressures of regional and global forces, all aggravated by climate change.

Conserving biodiversity

Biodiversity can be thought of as the information content of living systems that has accumulated over evolutionary time as organisms have adapted to their physical environment and to one another, both within species for purposes of competition for resources and cooperation in breeding, and between species in terms of predation and disease. A simple classification divides biodiversity into genetic, species, and ecosystem diversity. The species level is often used as a proxy for total biodiversity, for which there are two key indicators of relative abundance: species richness (the number of different species in an area); and endemism (the proportion of those species that occur nowhere else). The ecosystem level aggregates communities of organisms into 'kinds' of living system (grassland, woodland, sea-bed, etc.), but these are rather arbitrary since they can be sub-divided almost indefinitely into smaller and smaller sub-types, or aggregated almost seamlessly up to and including the planetary ecosystem or biosphere. Biodiversity loss is the simplification of living systems through the deletion of lineages, species and representatives of other taxa, as a result of such causes as over-harvesting, clearance for other land uses, pollution, and a general lack of care for the web of life when its survival is in competition with other human priorities. The process also erodes the integrity of the ecosystem concerned and is generally accompanied by the economic, cultural, and spiritual impoverishment of the human communities that formerly depended on or were inspired by those species and ecosystems. Two practical, net consequences of biodiversity loss are to reduce the resilience of local ecosystems and peoples, and to starve us of options for increasing resilience elsewhere, thus degrading our capacity to adapt to climate change.

Tracking investments in climate, desertification, and biodiversity

All aid investments are points of entry into systems that are too complex to understand in full. But to make sense of them enough so that we ourselves, as well as bureaucrats, politicians, and accountants, can follow what's going on, and so that journalists can explain it to the voters who are being taxed to pay for it, we have to classify our actions. As a start, we can look at clusters of investments to do with mitigating and adapting to climate change, combating desertification, and conserving biodiversity. The first two clusters are fairly straightforward, since they concern an identifiable issue – biosphere heating and climate change caused by the excessive exhaust gases produced by our civilisation, and reducing those emissions (mitigation) or managing their harmful consequences (adaptation), while trying to keep human development going despite them. Conserving biodiversity is much more all-embracing, bearing in mind that, according to the CBD, biodiversity includes everything alive and everything that organisms make, do, and is made from them, including ecosystems at all levels, species, and genetic information. The CBD seeks to guide us on how we use biodiversity so that it, and we, can keep human development under way. Finally, combating desertification is an investment cluster that straddles biodiversity and adaptation in dry-land circumstances, where human development must be sustained under particularly harsh and, with climate change, often worsening conditions, by conserving ecosystems and managing water at the margins of the habitable world. It is the concept of 'keeping human development going' (also known as 'sustainable development') that unites the issues of mitigation, adaptation, desertification, and biodiversity, and also the purposes and performance of overseas aid.

International agreements on mitigation, adaptation, desertification, and biodiversity, and groups with a special interest in these subjects, have steered the donor agencies towards keeping track of investments in those particular areas, so that they can to some extent be distinguished from all others. This has involved in each case a long process of research and dialogue to define, and eventually to agree, what is meant by an investment that should be recorded as addressing these challenges. The resulting check-lists represent kinds of investment that qualify to be recorded, and are a reasonable way to classify the flow of aid funding into the broad areas concerned. In the case of mitigation and adaptation, the check-lists are known as the Rio Climate Markers (Box 3.4), and were agreed in 1998 for mitigation, and in 2009 for adaptation. These gave rise to lists of specific kinds of action that can be counted as contributing to mitigation (Box 3.5) and adaptation (Box 3.6). For biodiversity and desertification, the starting points were the other Rio treaties, the CBD and UNCCD, which yield lists of actions relevant to their implementation (Boxes 3.7 and 3.8). I must stress that the contents of Boxes 3.5–3.8 are summaries and do not reproduce the relevant treaty texts, so they should not be used for any purpose in which the texts (in which the placing of a single comma can be vital) are important.

Box 3.4 The Rio Climate Markers

The OECD Development Assistance Committee (DAC) gathers statistics on aid flows, and asks donors to use 'markers' to classify their aid according to its relevance to the aims of the United Nations Framework Convention on Climate Change (UNFCCC), which was signed at Rio de Janeiro in 1992. The first Rio Climate Marker was agreed in 1998 and was for mitigation, but in 2009 the DAC also approved a marker to track aid in support of adaptation. Both markers allow for projects and their budgets to be recorded as 'principally' focused on climate change (i.e. that it is the main reason for the investment) or only 'significantly' so (i.e. that the investments have other primary aims, but they can also help with mitigation and/or adaptation). The idea is that it should be possible to report that a particular share of total aid was spent on mitigation or adaptation during a particular period, that some donors were spending more than others, and that the total share was increasing or decreasing over time, allowing donors to answer questions on what they have been doing to meet the climate challenge. The numbers thus generated are only approximations, but they do shed light on trends and they refer to useful check-lists of kinds of projects that might qualify as actions that could contribute to mitigation or adaptation. It is these check-lists that are used to classify the objects of evaluation here, when it comes to assessing the likely performance of mitigation and adaptation investments from the point of view of their relevance, effectiveness (tactical or short-term effects), and impact (strategic or long-term effects).

Source: OECD (2011).

Box 3.5 Actions that promote reduction of GHG emissions

Practical actions for climate change mitigation:

- *Technology*. Reducing or stabilising GHG emissions in the waste and sewage management, transport, energy, agricultural, construction, industrial, and other sectors through application of new and renewable forms of energy, measures to improve the energy efficiency of existing generators, machines and equipment, or demand-side management. See Chapter 4.
- *Ecology*. Protecting or enhancing GHG sinks and reservoirs through forest protection, avoided deforestation, sustainable forest management, reforestation, restoration of disturbed ecosystems (including soils through organic farming), rehabilitation of areas affected by drought and desertification, and sustainable management and conservation of oceans and other marine and coastal ecosystems, wetlands, wilderness areas, and other ecosystems. See Chapter 5.
- *Capacity*. Developing, transferring, and promoting emission-reducing technologies and know-how, including building capacity to control, reduce, prevent or reverse emissions of GHGs in the waste and sewage management, transport, energy, agricultural, forestry, construction, industrial, and other sectors. See Chapters 4 and 5.

Enabling frameworks for climate change mitigation:

- *Mainstreaming*. Integrating mitigation concerns and priorities within development processes, through preparation of national inventories of

GHGs (emissions by sources and removals by sinks), mitigation-related policies, legislation, and economic analysis and instruments, low-carbon development strategies and plans, needs assessments on mitigation technology, and building relevant institutional capacity. See Chapters 4 and 5.

- *Regulations and incentives.* Strengthening regulatory frameworks to put in place fiscal, economic, legal, and other incentives that discourage GHG emissions and encourage investment in reducing them. See Chapters 4 and 5.
- *Education and training.* Promoting mitigation-related education, training, and public awareness. See Chapters 4 and 5.
- *Research and monitoring.* Promoting research and monitoring efforts focused on mitigation and the understanding of oceanographic and atmospheric systems and processes. See Chapters 4 and 5.

Sources: OECD (2011); Caldecott *et al.* (2014, Annex 2: 76).

Box 3.6 Actions that promote adaptation to climate change

Practical actions for climate change adaptation:

- *Resilience.* Making landscapes, farming systems, and communities more resilient to environmental change, including (as appropriate to changes anticipated in each location) through measures to safeguard or restore the ecosystem services of water catchments, floodplains, wetlands, mangroves, coral reefs, beach dunes, and aquifer recharge areas, conserving water and introducing water-saving irrigation methods, introducing crops that are resistant to heat, drought, submergence and salinity, prophylaxis against vector-borne and other diseases, amending fishery management practices in response to new ecological conditions and changing fish populations, promoting diverse forest management practices and species, developing emergency prevention and disaster preparedness measures (including insurance and engineering works to relieve known threats, e.g. from glacial lake outburst floods and sea-borne storms). See Chapter 6.
- *Knowledge.* Promoting stakeholder environmental monitoring and networking to enhance sharing of knowledge on environmental change, threats, solutions, and adaptation best practices (as appropriate to changes anticipated in each location), including the building of social capital, cooperation and adaptation/disaster preparedness, and the production and dissemination of public information materials on the principles and practices of adaptation. See Chapter 6.

Enabling frameworks for climate change adaptation:

- *Mainstreaming.* Supporting the integration of adaptation into national and international policies, plans, and programmes, and strengthening the capacity of key national institutions (including finance and planning ministries) to coordinate and plan for adaptation activities, and integrate adaptation into planning and budgeting. See Chapter 6.
- *Disaster risk reduction.* Building capacity for disaster risk reduction, preparation, and management at local, national, and regional level, by making disaster-relevant information and tools more accessible to all, by promoting disaster consciousness in adaptation policies, strategies, and programmes, and encouraging systematic dialogue, information exchange, and joint working between climate change and disaster reduction institutions and experts, in collaboration with policy makers and development practitioners. See Chapter 6.
- *Education and training.* Promoting adaptation-related education, training, and public awareness-raising. See Chapter 6.
- *Research and monitoring.* Promoting research focused on environmental change, and weather, climate and water monitoring and information systems, including observation and forecasting, impact, and vulnerability assessments and early warning systems, and on how to make landscapes, farming systems, and communities more resilient to detected or anticipated changes. See Chapter 6.

Sources: OECD (2011); Caldecott *et al.* (2014: 76–77).

Box 3.7 Actions for implementing the Convention on Biological Diversity (CBD)

Practical actions for biodiversity conservation:

- *Protection of ecosystems and species.* Creating and managing protected areas. Preserving endangered species, including reintroductions. Suppressing poaching and illegal trade in endangered species. Developing legislative frameworks for protecting species and sites. Supporting sustainable development in the buffer zones of protected areas. Building capacity of authorities and personnel in charge of protected areas. Research in support of protected area management. Eradicating and resisting the spread of invasive species. Establishing and operating *ex-situ* (off-site) conservation facilities (zoos, botanical gardens and conservatories, seed banks, gene banks, etc.). Research in support of *ex-situ* conservation.

- *Water and sanitation*. Protecting water resources and improving water quality where beneficial effects on fauna and flora are expected. Preventing and ameliorating water pollution, particularly in rivers. Promoting the integrated management and protection of water catchments. Promoting the integrated management and protection of wetlands.
- *Agriculture*. Reducing land use change and expansion of agriculture at the expense of natural areas. Promoting environment-friendly agricultural practices, including agroforestry and organic farming. Promoting farming practices that limit chemical inputs (fertilisers, pesticides), or that use integrated pest management. Promoting soil conservation and resisting soil erosion.
- *Forests*. Preserving forests. Promoting sustainable forest management (in particular, through community forestry). Suppressing illegal logging and resisting deforestation for agriculture or livestock. Rehabilitating areas affected by desertification. Promoting natural forest management and multi-species plantations through forestry and agroforestry in areas with high biodiversity value. Institutional support, raising awareness.
- *Harvesting (fishing, hunting, and gathering)*. Promoting sustainable fisheries in marine, coastal, and lake/river environments. Suppressing overfishing, illegal fishing, and destructive fishing practices. Conducting inventories and research on the status of stocks, fishing capacity. Preserving nursery areas, including mangroves. Preserving coral reefs. Promoting sustainable hunting practices and anti-poaching efforts. Promoting sustainable harvesting, including non-timber forest products. Developing the legislative framework for fishing, hunting, and gathering. Institutional support, raising awareness.
- *Business and economy*. Suppressing illegal trade in endangered species, including raising awareness among the authorities and consumers, training of personnel (customs etc.). Suppressing illegal trade in animal products/plants (bush meat, ivory, orchids, etc.). Preventing the importation of invasive species. Promoting sustainable ecotourism, vision tourism, tourism valuing biodiversity. Promoting research on the economic valuation of biodiversity. Establishing mechanisms for equitable access and benefit sharing of genetic resources. Institutional support, raising awareness.

Enabling frameworks for biodiversity conservation:

- *Mainstreaming*. Preparing biodiversity inventories. Preparing and implementing national plans and strategies for biodiversity conservation. Developing legislative frameworks relating to species and ecosystems, including the issue of invasive species. Mainstreaming biodiversity in national strategic frameworks and local policies on land use. Developing innovative, long-term, financing mechanisms (trust funds, ecological taxes, etc.). Capacity building. Developing tools for decision support (databases, information systems).

- *Education, training and research.* Training, research and teaching in taxonomy, biodiversity, ecology, etc. Institutional support, capacity-building, awareness-raising and public information on biodiversity in relation to forests, harvesting, business and economy, farming, health, etc.

Source: Fiche 1B on key sectors and activities related to the CBD, in EuropeAid (2010).

Box 3.8 Actions for implementing the Convention to Combat Desertification (UNCCD)

Practical actions for combating desertification:

- *Water.* Promoting sustainable water management/integrated water management. Promoting sustainable management and protection of catchment areas. Promoting integrated management of international rivers, lakes, wetlands and aquifers.
- *Agriculture.* Promoting sustainable irrigation. Resisting soil salinisation and promoting the rehabilitation of salinised areas. Promoting sustainable management of grasslands. Promoting water-efficient agricultural practices. Promoting farming practices that protect soils (agroforestry, 'cover crops', etc.). Promoting anti-erosion works in rural areas (hedges, walls, terraces, etc.). Strengthening agricultural development services, farmers' organisations.
- *Forest and natural areas.* Promoting the protection and sustainable management of dry forests. Establishing protected areas and promoting sustainable management of natural areas. Promoting the sustainable management of forests for fuelwood, alternative energies, and energy efficiency at domestic level (improved ovens etc.). Promoting afforestation (planting new forests) and reforestation (re-planting lost forests). Promoting sand dune fixation.

Enabling frameworks for combating desertification:

- *Institutional support.* Supporting hydrological and meteorological services. Developing legislative and institutional frameworks (including on land tenure). Supporting the preparation and implementation of National Action Programmes to combat desertification. Integrating desertification issues in national strategic frameworks and policies on land use.

- *Training, research, and teaching.* In agriculture, water resources management, farming practices, etc. Developing environmental technologies to combat desertification. Developing databases, information systems on desertification.
- *Raising public awareness.* Building public awareness on desertification-related issues.
- *Strengthening civil society.* Building the capacity of civil society (community, charitable, voluntary, and private-sector institutions) on desertification-related issues.

Source: Fiche 3B on key sectors and activities related to the UNCCD, in EuropeAid (2010).

The check-lists of actions in Boxes 3.5–3.8 have in common that they divide tasks into two main categories:

- *actions*, based on a theory of change that says something like 'if we do *this* then it should result in a certain amount of *that*', and marked by descriptions such as protecting, reducing, managing, planting, or making things happen; and
- *processes*, based on a theory of change that says something like 'if we improve people's knowledge or relationships then their behaviour will improve', and marked by descriptions such as building capacity, raising awareness, developing policies, integrating concerns, or mainstreaming.

It is clear that both are important, since without the *processes* the *actions* will occur in a donor-driven vacuum, with zero sustainability. Some observers are drawn by the 'value for money' argument towards valuing only tactical effectiveness (i.e. through action), but others argue that unless people care, or understand enough about how it works and what's at stake to care, then any specific action will have an impact only to the extent that it also educates (i.e. through process). This divergence between a short-term, instrumental attitude, and a long-term, educational one has always been a feature of overseas aid discussions. In any case, all three treaties wisely opt for both approaches to all four themes.

Another notable feature of Boxes 3.5–3.8 is the degree of overlap among the different sets of actions and priorities. The technology of mitigation is distinct, but the ecology of mitigation largely concerns finding ways to encourage and enable people to conserve ecosystems as carbon stores, even though those same ecosystems will also be 'storing' biodiversity and providing many other goods and services. These latter may be seen as 'co-benefits' from a mitigation point of view, while carbon storage might be seen as a 'co-benefit' from a biodiversity, water catchment, or adaptation point of view. Similarly, many of the actions needed for adaptation (e.g. under 'resilience' in Box 3.6) are identical to those required in combating desertification or conserving biodiversity, with the details depending largely on ecological context. Moreover, for mitigation, burning living or recently dead ecosystems (i.e. components of biodiversity at the ecosystem level) is nearly as important a source of GHGs as burning long-dead ones (i.e. fossil fuels), and if left alone or restored, living ecosystems can absorb carbon from the air and store it safely.

Meanwhile, for adaptation, living ecosystems can help regulate water supplies, protect soils from depletion and erosion, absorb the energies of storm impacts, and perform other services that we can rely on to take the edge off some of the effects of climate change. And biodiversity in all its forms (of ecosystems and ecosystem goods and services, including species and genetic resources) also underpins the process of ecosystem management known as agriculture, which when adapted to harsh conditions can make the difference between a desertified and a verdant landscape. These inter-connections pose a problem for those trying to keep track of aid investments and working out what effect can be attributed to any particular activity or budget line. In the chapters that follow, which explore what is involved in spending money in line with the various thematic markers, the solution used is to distinguish between investments that specifically promote the mitigation of climate change (Chapters 4 and 5), and those that promote adaptation and resilience to climate change at local, regional and global level, and at all time scales (Chapter 6).

A word of caution

Sustainability and mass extinction

Humanity's efforts to promote sustainability in recent decades have not been very effective in reducing overall human impacts on the biosphere. Structural biases in human attention, on average and at an institutional level, seem to inhibit long-term and systemic thinking, and paying attention to equity between and within human groups and generations, or between human and non-human stakeholders. Meanwhile, the human system has grown in scale and diversified in content, driven by inventiveness, ingenuity, acquisitiveness, reproduction, and the quest for wealth. Although rooted in all the events of the Holocene (the last 11,700 years or so), a take-off point was reached in about 1950, which is now known as the Great Acceleration, and marks the start of the Anthropocene that reflects the influence that humanity is having on nature (Waters *et al.*, 2016). Our species is now producing about 50 billion tonnes of GHGs (measured as gigatonnes of carbon dioxide equivalent, or $GtCO_2e$) every year (Flannery,

2015). The beginnings of inevitable climate change are becoming awful for hundreds of millions of people, in various ways both direct and indirect, and this will worsen whatever we do now (IPCC, 2014; Wadhams, 2016). Meanwhile, we continue to take over and finish off for our own purposes all those natural ecosystems that are not legally protected, and damage or destroy many of those that are, with the result that up to a million species a year are becoming committed to extinction (Box 3.9), and biodiversity is declining fast by all available objective measures (e.g. UNEP, 2012c; WWF, 2014, 2016).

Box 3.9 Global species richness and extinction rates

Estimates of the number of species in the world range up to about 15 million based on those that are already known to taxonomic science plus the rate of new discoveries, but these estimates greatly under-represent little-known groups such as sea-bed nematodes, tropical invertebrates (notably beetles), and major groups of microorganisms. Faced with this degree of uncertainty, I am inclined to take 25 ± 15 million to represent the world's pre-industrial species richness, while accepting that most of those species are completely unknown. That most of them will soon be extinct is strongly suggested by the fact that the species we know about are not evenly distributed, and 34 biodiversity 'hotspots' have been identified where high levels of species richness and endemism and high rates of habitat damage combine to make their survival critical for biodiversity conservation. Between them, these hotspots contain most (60–80 per cent) of the world's species – at least 150,000 endemic plants, almost 12,000 endemic vertebrates, and millions of invertebrates – in what was originally about 15.7 per cent of the planetary ecosystem's land area. But more than 86 per cent of the total habitat in these hotspots has already been destroyed, and the destruction and fragmentation of such a large share of their ecosystems implies that at least half of the species originally present are in the process of going extinct. Not necessarily at once, but committed to extinction due to reduced populations and habitat loss, and such factors as the deaths of co-evolved partner species (pollinators, seed-dispersers, hosts, etc.).

So if people immediately stopped expanding their farms and ranches, draining wetlands, polluting reefs, setting fires, digging mines, logging forests, altering microclimates, etc., we would still be looking at well in excess of ten million species dead or dying, or up to about a million a year. The precise number will never be known and is not worth arguing about, but droughts, fires, sea-level rise and other effects of climate change are likely to be pushing extinction rates rapidly upward. Species extinction also occurred before humanity rose to power, but the current extinction rate is far higher than that prevailing in the past, at least during the intervals between 'mass extinctions'. These are intense bouts of worldwide extinction, of which five are known from the fossil record over the last 540 million years. The current rate of loss qualifies the present period (the 'Anthropocene') as the sixth mass extinction in the history of life on Earth, and the first to be caused by a single species, people. Thus the biosphere is now entering a phase in which large numbers of species will go extinct in the near future, along with many lineages, sub-species, genera and families. Whether it can be stopped is open to question, but it is important that we understand and learn from what we have done to nature.

Sources: Myers (1988, 1990); Myers *et al.* (2000); Groombridge & Jenkins (2002); CEPF (2015).

Changing priorities for ODA eligibility

Delivering overseas aid is becoming harder to do in many places, at least in the sense of people working in countries other than their own, paid for by institutions with funds raised elsewhere. This is partly because many countries have become too wealthy to qualify for overseas aid, as measured by GDP-based standards. From the point of view of relieving poverty, this can leave some countries in an anomalous position, with high GDP per person on average but with inequality that leaves large numbers of people in poverty. At the level of journalism, public opinion, and therefore politics in the OECD countries, it inevitably becomes hard to justify spending public funds on poor people, no matter how deprived, who are citizens of countries with, for example, space and nuclear weapons programmes (an obvious case being India). In such places, ODA tends to be replaced by charity and, in response to natural disasters that can happen to anyone, temporary humanitarian assistance. With poverty at the national level as the key yardstick for determining aid eligibility, the main aid donors naturally turn their attention elsewhere, so there are fewer countries eligible for aid, and it is harder to make progress in those that remain because of intractable social conditions.

From the point of view of environmental sustainability, however, things look very different. Many countries have become richer in terms of GDP but a lot poorer in natural ecosystems, which have been systematically logged, burned, mined, drained, dredged, farmed, or built on. Consequences include the closing off of development pathways that could have used those lost resources, which include biodiversity, and this is a fact which requires some learning, imagination, and a long-term perspective to appreciate. They also include the emergence of a new class of environmental problems, of which among the easiest to perceive are those to do with water – specifically, the regular supply of fresh, clean water. Here there are plenty of things that can be done, mostly involving people organising themselves to prevent fresh water from being polluted or privatised and charged for at a rate that is unaffordable for poor people, and to ensure the protection or restoration of catchment ecosystems and the recharge of aquifers.

Such efforts have a pro-poor orientation, which is attractive to anti-poverty donors, but also a pro-equity direction that may make it hard for governments to engage with. Besides, official donors may take the view that a wealthy country with internal problems of water equity should deal with its own issues through its own political processes. We are left with innumerable, smallish (but locally very important) water-ecology-equity problems around the world that are hard to address through traditional ODA mechanisms, unless they can somehow be aggregated and related to an over-arching challenge that is big enough to prompt government cooperation. Climate change adaptation is just such a challenge, and most countries are now willing to cooperate with one another in spending money on it at an annual rate similar to that already committed to ODA. The implication is that adaptation is providing an agreed substitute for poverty eradication as the main driver of overseas aid, although of course the two agendas have many areas of overlap, as they do with what are generally seen as the much lower priorities of promoting equity and halting biodiversity loss.

Meanwhile, the relationship between mitigation and adaptation is a complex and political one. It is complex because without effective mitigation to prevent runaway climate change, adaptation may well be impossible, and many of the things involved in mitigation also have consequences for resilience and adaptation, especially where they involve managing ecosystems, and *vice versa*. It is political because decarbonising the world's economy to an effective extent is so costly that few public leaders have been able to sell the idea to a stable majority of the populations that they represent, and many have anyway rejected or delayed any such initiative. Even fewer have been brave enough to use their executive authority to do what needs to be done in the face of political competition and a fearful population. And poorer countries are disinclined to accept what may be seen as instructions on mitigation from richer ones, without most of the costs being met by the latter, whose leaders are reluctant to make such commitments. In these circumstances, pending some electrifying historical event that will focus collective minds on the threat, we can only proceed pragmatically, imaginatively, and as best we can. Opinions on the likely outcome of our efforts are mixed, ranging from cautiously optimistic (Flannery, 2015), to resigned (Scranton, 2015). The latter summarises the position thus: "The problem with our response to climate change isn't a problem with passing the right laws or finding the right price for carbon or changing people's minds or raising awareness. *Everybody already knows*. The problem is that the problem is too big. The problem is that different people want different things. The problem is that nobody has real answers. The problem is that the problem is us" (*ibid.*: 68).

Terrorism

There is another dynamic at work that owes little directly to poverty or environmental issues, although both can be contributory factors, and that seems to negate the values on which the European Consensus on Development and the

United Nations' MDGs and Sustainable Development Goals (SDGs) are based. Unresolved ideological and/or specifically religious disputes seem to have multiplied in recent years, and there is an increasing willingness by extremists to resort to the tactics of terrorism. This is a big subject and a big problem, with large areas of the world now subject to chronic low-level violence that is often directed at the general public when it does not target those that are seen as representatives of 'enemy' groups, including aid workers. This makes it hard to deliver aid safely in many places, whether focused on health, poverty, biodiversity, or climate change. But the key problem is that no conceivable combination of wealth, equity, and sustainability can be *guaranteed* to demotivate all those committed to the practice of terrorism, at least in the short term. Various solutions are being tried, including military strikes, preventative security, counter-propaganda, and re-education, but these often address symptoms rather than root causes. In the absence of other ideas and without guarantees of success, we will just have to try to build wealth, equity, and sustainability as widely as we can, while also trying to settle fairly, comprehensively, and in line with our values, as many of the underlying disputes as possible.

Nationalism

A love of homeland can motivate care and attention to the ecological and social integrity of local environments by their inhabitants. In the form of nationalism, however, this emotion can be exploited by deliberately fanning a fear of outsiders among the public and using it to advocate policies that reduce cooperation among peoples and countries. The June 2016 referendum in the UK on its membership of the EU is a case in point, in view of the overtly nationalist themes of the case being made to leave, which largely focused on fear of immigration and the loss of national control over legislation. Regardless of the details of the conduct of the referendum, and the ways in which competing visions for the future of the country were presented, the result was narrowly in favour of the UK leaving the EU. Efforts to offset the effects of this outcome through negotiated rearrangements in all sectors are currently under way, with unpredictable implications for overseas aid and the climate change response. This disruption of international cooperation when it is most needed against climate change is likely to be very harmful. It is also indicative of a trend towards nationalist (or anti-internationalist) activism among political movements in numerous countries, which is deeply unhelpful in present global circumstances.

References

Caldecott, J. O. (2008) *Water: the Causes, Costs and Future of a Global Crisis*. Virgin Books (London).

Caldecott, J. O., Halonen, M., Sørensen, S. E., Dugersuren, S., Tommila, P. & Pathan, A. (2010) *Evaluation of Sustainability in Poverty Reduction: Synthesis*. Evaluation Report 2010: 4. Ministry for Foreign Affairs (Helsinki).

Caldecott, J. O., Dornau, R., Gollan, J., Halonen, M., Saalismaa, N., Schilli, A., Simonett, O., Stuhlberger, C. & Tommila, P. (2014) *Technical Report on Effectiveness of the Swiss International Cooperation in Climate Change Mitigation and Adaptation Interventions 2000–2012*. Mandate of the Swiss Agency for Development and Cooperation (SDC) and the State Secretariat for Economic Affairs (SECO) (Bern).

CEPF (2015) *The Biodiversity Hotspots*. Critical Ecosystem Partnership Fund. www.cepf.net/resources/hotspots/Pages/default.aspx (consulted 13 Jan 2016).

Chambers, R. (1983) *Rural Development: Putting the Last First*. Prentice Hall (Upper Saddle River, New Jersey).

Clover, C. (2004) *The End of the Line*. Ebury Press (London).

Easterly, W. (2006) *The White Man's Burden: Why the West's Efforts to Aid the Rest Have Done So Much Ill and So Little Good*. Penguin (London).

EuropeAid (2010) *The Multilateral Environmental Agreements and the Rio Markers: Information Note*. EuropeAid Unit E6 ('Natural Resources'). European Commission (Brussels).

Flannery, T. (2015) *Atmosphere of Hope: Solutions to the Climate Crisis*. Penguin (London).

Groombridge, B. & Jenkins, M. D. (2002) *World Atlas of Biodiversity: Earth's Living Resources in the 21st Century*. UNEP Word Conservation Monitoring Centre and University of California Press (Berkeley, California).

Hardin, G. (1993) *Living Within Limits: Ecology, Economics and Population Taboos*. Oxford University Press (New York).

Houghton, J. (2004) *Global Warming: the Complete Briefing*. Cambridge University Press (Cambridge).

IPCC (2014) *Climate Change 2014: Synthesis Report. Contribution of Working Groups I, II and III to the Fifth Assessment Report of the Intergovernmental Panel on Climate Change* (edited by the Core Writing Team, R. K. Pachauri & L. A. Meyer). Intergovernmental Panel on Climate Change (Geneva).

Lovelock, J. (2006) *The Revenge of Gaia*. Allen Lane (London).

MA (2005a) *Living Beyond Our Means: Natural Assets and Human Well-being*. Millennium Ecosystem Assessment. Island Press (Washington, DC).

MA (2005b) *Ecosystems and Human Well-Being: Current State and Trends*. Millennium Ecosystem Assessment. Findings of the Condition and Trends Working Group. Island Press (Washington, DC).

McMichael, A. J., Campbell-Lendrum, D. H., Corvalán, C. F., Ebi, K. L., Githeko, A. K., Scheraga, J. D. & Woodward, A. (editors) (2003) *Climate Change and Human Health: Risks and Responses*. World Health Organization (Geneva).

Meadows, D. H., Meadows, D. L., Randers, J. & Behrens, W. W. (1972) *The Limits to Growth*. Universe Books (New York).

Meadows, D. H., Meadows, D. L. & Randers, J. (1992) *Beyond the Limits*. Chelsea Green (White River Junction, Vermont).

Monbiot, G. (2006) *Heat: How to Stop the Planet Burning*. Allen Lane (London).

Moyo, D. (2009) *Dead Aid: Why Aid is Not Working and How there is Another Way for Africa*. Allen Lane (London).

Myers, N. (1988) Threatened Biotas: "Hot-spots" in Tropical Forests. *The Environmentalist* **8**(3): 187–208.

Myers, N. (1990) The Biodiversity Challenge: Expanded Hot-spots Analysis. *The Environmentalist* **10**(4): 243–256.

Myers, N., Mittermeier, R. A., Mittermeier, C. G., da Fonseca, G. A. B. & Kent, J. (2000)

Biodiversity Hotspots for Conservation Priorities. *Nature* **403**: 853–858.
Myers, S. S. (2009) *Global Environmental Change: the Threat to Human Health*. Worldwatch Report 181. Worldwatch Institute (Washington, DC).
OECD (2011) *Handbook on the OECD-DAC Climate Markers*. OECD (Paris).
Oxfam (2007) *Up in Smoke? Asia and the Pacific: the Threat from Climate Change to Human Development and the Environment*. The fifth report from the Working Group on Climate Change and Development. Oxfam (London).
Pearce, F. (2006) *When the Rivers Run Dry*. Eden Project/Transworld (London).
Scranton, R. (2015) *Learning to Die in the Anthropocene: Reflections on the End of a Civilization*. City Lights (San Francisco).
UNDP (2015) *Human Development Report 2015: Work for Human Development*. United Nations Development Programme (New York).
UNEP (2012a) *Fifth Global Environment Outlook (GEO-5): Chapter 6 – Chemicals and Waste*. United Nations Environment Programme (Nairobi).
UNEP (2012b) *Fifth Global Environment Outlook (GEO-5): Chapter 4 – Water*. United Nations Environment Programme (Nairobi).
UNEP (2012c) *Fifth Global Environment Outlook (GEO-5): Chapter 5 – Biodiversity*. United Nations Environment Programme (Nairobi).
Wadhams, P. (2016) *A Farewell to Ice*. Allen Lane (London).
Waters, C. N., Zalasiewicz, J., Summerhayes, C., Barnosky, A. D., Poirier, C., Gałuszka, A., Cearreta, A., Edgeworth, M., Ellis, E. C., Ellis, M., Jeandel, C., Leinfelder, R., McNeill, J. R., Richter, D. deB., Steffen, W., Syvitski, J., Vidas, D., Wagreich, M., Williams, M., An, Z.-S., Grinevald, J., Odada, E., Oreskes, N. & Wolfe, A. P. (2016) The Anthropocene is Functionally and Stratigraphically Distinct from the Holocene. *Science* **351**(6269): 138–141.
WWF (2014) *Living Planet Report 2014: Species and Spaces, People and Places* (edited by R. McLellan, L. Iyengar, B. Jeffries, & N. Oerlemans). WWF (Gland).
WWF (2016) *Living Planet Report 2016: Risk and Resilience in a New Era*. WWF International (Gland).

Chapter 4

Technological mitigation

Substitution and efficiency investments

Overview

Substitution investments in my sample aimed to minimise release of GHGs per unit of energy at every stage of economic activity, including power generation, transmission, and end use. For generation, relevant technologies minimise emissions at source, using wind (on-shore, off-shore), solar (thermal, electric, domestic, industrial), tidal (barrage, lagoon, sea-bed turbine), wave, hydroelectricity (micro, big dam, run-of-river), geothermal (domestic, industrial), and biomass (biogas, biofuels). For transmission, the aim is to minimise energy loss in power lines and transformers, while technologies that target end use aim to minimise energy wastage in obsolete equipment and construction materials, poorly designed buildings, etc., and can also address urban air quality and transport issues. In evaluation terms, there is a strong emphasis on effectiveness, with evidence taking the form of expected or documented emission reductions per unit of technology, time, and money.

Hydroelectricity

Although large dams use large amounts of concrete to build, the manufacture of which is a major GHG emission source, and have other drawbacks (e.g. construction emissions, replacement of natural ecosystems by the dam lake, earthquakes caused by the weight of the dam lake, siltation of the dam lake especially where catchment ecosystems are degraded, loss of land by indigenous peoples, and disruption of wildlife migrations and dispersion), in the long term they can have the net effect of reducing GHG emissions when compared with other ways of generating electricity. These gains can be maximised, however, by concentrating on rehabilitating hydropower systems in which the major sunk costs of construction have already occurred (Box 4.1), or else through promoting run-of-river and/or small-scale hydropower in the context of decentralised electricity systems (Box 4.2).

Box 4.1 Rehabilitating the Drin River Cascade in Albania

The project helped to rehabilitate four large hydropower plants on the Drin and Mat rivers in Albania. The priority was to improve energy production and reliability, access and energy security by prolonging the life of the utilities, and improving dam safety and the use of water. In the context of support by several donors, a Swiss grant focused on delivering hydromechanical equipment for the Fierza hydropower plant (HPP) at the top of the cascade. The investment improved reliability and reduced power cuts, even during extreme floods, increased efficiency by 3–4 per cent, and extended the lifespan of the Fierza HPP. A field visit also revealed that the investment and its follow-up activities had helped to increase dam safety. Reducing GHG emissions was not an explicit goal of the investment, but it nevertheless occurred due to a reduced need to import electricity from neighbouring countries with higher emission rates per unit of power generated, and a reduced use of diesel generators during power cuts, so the investment was given a moderate mitigation effectiveness score of 4.

Sources: Appendix 1 (37); Caldecott *et al.* (2014: 204–206, project UZ-00574).

Box 4.2 Hydropower investments

"The World Bank Group's [WBG's] direct lending for renewable energy is dominated by hydropower, the only grid technology for which there is a substantial evaluable record at the WBG. Among evaluated hydropower plants, 76 per cent had outcomes rated as moderately satisfactory or better, with better ratings in recently initiated projects. Unsuccessful projects are often those for which preparation or implementation of resettlement plans has been ineffective. About two-thirds of hydropower investment volume now goes to run-of-river hydropower (that is, without substantial reservoirs), which has less potential for local social and environmental damage but is more vulnerable to climate change" (Caldecott *et al.*, 2014: 406).

Source: Appendix 1 (45).

Power system rehabilitation, efficiency and sustainability

Among other terrible consequences, civil war and social chaos can delay investment in what could otherwise be a smooth transition from obsolete (i.e. low-efficiency, high-emission) power generation and transmission to more climate-friendly arrangements. This applied particularly during the 1990s and 2000s in the Balkans and parts of South-west Asia, including Macedonia, Serbia,

Tajikistan, Kosovo, Kyrgyzstan, and Albania. Once peace is restored, donors may focus on delivering an energy efficiency portfolio that targets the quality of power generation and transmission, largely through the replacement of obsolete or war-damaged equipment at power plants, and new control and transmission systems (Box 4.3). A additional aim may be to modernise managerial and financial (e.g. billing) arrangements to ensure that revenues are at least sufficient to maintain the power system in the future. The relevance to climate change will vary, however, according to the success of project implementation and how integrally climate impacts are planned for, reported and monitored, and these factors interact with the ways in which investment financing is made available (Box 4.4).

Box 4.3 Modernising the Nikola Tesla Thermal Power Plant B (TENT B1) in Serbia

The project designers assumed that the Serbian economy would continue to depend on coal for energy production, at least for a while, so they considered that immediate measures to reduce pollution were critical. The project initially helped improve energy efficiency and reliability at TENT B1, thereby reducing power cuts and emissions of CO_2 and other pollutants. Although emission reductions amounted to only about 2 per cent of TENT B1 emissions (or about 90,000 tCO_2/year), the project was scored 5 for mitigation effectiveness, partly because installing a new monitoring and control system also prepared the way for other investments (by the German development bank KfW and the European Agency for Reconstruction) in filters and other equipment for more efficient energy production and environmental protection. The project drew attention to trade-offs between achieving positive environmental and socio-economic impacts while extending the life-span of coal power, and highlighted the importance of understanding the overall context, and the role of assumptions and system boundaries, when making net assessments of impact.

Sources: Appendix 1 (38); Caldecott *et al.* (2014: 180–184, project UR-00269).

Box 4.4 Energy efficiency investments

"Phase I of the evaluation (World Bank, 2009) assessed the most important barrier-removing policies: energy price reform and promotion of energy efficiency policies such as building and appliance standards. It noted that the Bank had pursued price reforms in energy but had put few resources into energy efficiency, although attention to policy-efficiency linkages had since increased. Owners of factories and buildings often fail to borrow for apparently highly profitable energy efficiency opportunities. The WBG's diagnosis is that borrowers lack information, and lenders lack experience and comfort with energy efficiency project finance. The main WBG response has been to support financial intermediaries with guarantees and technical assistance. These programmes have been directed to China and Eastern Europe, where energy efficiency has long been low. Parallel programmes have been implemented by the World Bank and IFC [International Finance Corporation (of the World Bank Group)] (both supported by the GEF [Global Environment Facility]) but without much communication between them. Contrary to expectations, loan guarantees have turned out not to be a temporary, market-transforming measure that could be discontinued once the banks gained familiarity with energy efficiency lending. Inadequate lending for energy efficiency often reflects wider credit market failures, including onerous requirements for collateral. Because borrowers achieved high rates of return, guarantee programmes could achieve higher impact through tighter targeting on less creditworthy companies, such as SMEs [small and medium enterprises]. World Bank-supported projects have been successful in introducing energy service companies (ESCOs) to China, with high returns, significant GHG impacts, and spontaneous replication. Meanwhile, IFC's approach of screening its clients for energy efficiency opportunities leads mostly to small loans with low GHG impacts. Three areas of existing activity stand out as having high impact and high potential for scale-up, these being:

- proactive IFC support for energy efficiency in the atypical but important cases of large, carbon-intensive factories that face credit or information barriers;
- increased support for transmission and distribution loss reduction, which offers high economic rates of return and lifetime carbon returns of 7–15 kg/dollar; and
- substitution of compact fluorescent lamps for incandescent lamps offers very high estimated direct economic returns (in saved energy), together with deferred construction of power plants and emissions reductions of 27–134 kg/dollar."

Sources: Appendix 1 (45); World Bank (2010); quotation from Caldecott *et al.* (2014: 406–407).

Coal, biomass, biogas, and biofuels

Coal is the necromass of ancient ecosystems, in which atmospheric carbon has been captured and then fixed, first as plant material and then as deposits of elemental carbon isolated from the biosphere within geological formations. Once a certain technological stage is reached it is easy to dig up such carbon deposits and burn them, thus restoring the atmospheric carbon at a rate far higher than that of its original deposition, which is a major cause of our current problems. That coal deposits are so stable under these storage conditions has implications for modern negative-emission technologies, but meanwhile there is the challenge of transitioning away from coal as a significant energy source (Box 4.5).

Box 4.5 Coal power investments

"Coal is relatively cheap but also a major source of GHG emissions, and WBG support is now restricted (by its own *Strategic Framework on Development and Climate Change*, SFDCC) to cases where coal has the lowest cost after environmental externalities have been considered, there is optimal use of energy efficiency, and no concessional funds are available to finance the incremental cost of low-carbon alternatives. The IEG [Independent Evaluation Group (of the World Bank)] examined five pre-SFDCC coal power projects to determine whether WBG involvement contributed to greater efficiency and whether lower-carbon alternatives had been considered. It found that none of the investment cases would have met the SFDCC criteria, either because they were not least-cost for generation after accounting for local air pollution or because they did not fully explore efficiency alternatives. The complexity of the issues, however, is illustrated by IFC's support for a super-critical coal plant in India. It will be one of the largest point sources of CO_2 on the planet, but it may nevertheless have reduced emissions by about 10 per cent compared with a scenario without IFC involvement, and indirectly accelerated the diffusion of this higher-efficiency technology in a country that will continue to rely on coal for decades. More than a quarter of India's power is lost in transmission and distribution, however, and a nationwide reduction in distribution losses and other efficiency measures can offer higher returns in power availability, local environmental improvement, and GHG reductions than new construction" (Caldecott *et al.*, 2014: 407–408).

Source: Appendix 1 (45).

Obtaining energy from burning or fermenting biomass is said to be 'carbon-neutral' because it involves returning to the air carbon recently absorbed from it, rather than carbon that had been taken from the active biosphere in the distant past and hidden away as geological deposits (Box 4.6). At an industrial scale, serious drawbacks can include the large-scale replacement of natural ecosystems, such as rainforests by oil-palm plantations to produce 'biodiesel', or food-producing systems, such as when farmlands are converted to maize or sugar cane to produce 'bioethanol' (Box 4.7).

Box 4.6 Transforming the charcoal sector in Tanzania

The project was designed to explore the development of a sustainable charcoal trade in Tanzania, a country with many Village Land Forest Reserves that could be meeting urban demand for charcoal for the sustainable benefit of rural communities. Unfortunately, resistance by traders in illegal and unsustainable charcoal seems to have prevented implementation, so effectiveness was given a score of only 3.

Sources: Appendix 1 (37); Caldecott *et al.* (2014: 10, project 7F-07802).

Box 4.7 Sustainability standards for biofuels

The project tried to off-set the drawbacks of large-scale biofuel production by developing and promoting the adoption of principles and criteria for environmentally and socially sustainable production and trade in biofuels. This supported the Round-table on Sustainable Biofuels at a time (2008–2012) when huge controversy surrounded the consequences of US and EU efforts to require an increased proportion of bioethanol and biodiesel in gasoline and fuel oil. In this context the idea of exploring ways to make production and trade in biofuels more sustainable was potentially useful and may have contributed to policy reviews by the EU and US, and the deliberations of the Round-table on Sustainable Palm Oil and its Indonesian equivalent, so the investment was given a moderate mitigation effectiveness score of 4.

Sources: Appendix 1 (37); Caldecott *et al.* (2014: 9–10, project UR-00339).

Mixed renewables

The most appropriate mix of energy sources may be chosen from a menu of potential low-carbon technology investments for deploying in a location according to its special ecological and social conditions, ideally in the context of an over-arching low-carbon development plan. Such plans should extend their vision far beyond the technology and investment environment affecting mitigation technologies, to include an integrated and economy-wide approach covering low-emission and climate-resilient development in a whole territory, for example as devised by the Indonesian provinces of Papua (Suebu, 2009), East Kalimantan (Stafford, 2009) and Jambi (DNPI & OGJ, 2010). Making a location-specific selection can be supplemented by approaching the issue from the other direction, by raising awareness among donors of what the options might be, and how to choose them (Boxes 4.8 and 4.9).

Box 4.8 The Renewable Energy Programme in Vietnam and elsewhere

Several Swiss contributions to World Bank and International Finance Corporation projects focused on facilitating investments in unspecified renewables (small hydro, wind, biomass, etc.), focused on: (a) deploying technical assistance in Vietnam, which was scored 5 for mitigation effectiveness because of anticipated synergies with cleaner production initiatives in that country; (b) advising on four pilot projects in different renewable energy sectors, which was scored 4; and (c) a more strategic investment in 'Scaling Up Renewable Energy Program in Low Income Countries', to help show that developing a renewable energy supply is feasible and beneficial in low-income countries, by supporting them in expanding energy access through scaled-up renewables deployment, and by triggering change in the renewables market through government support for market creation, private-sector implementation, and productive energy use, which was scored 5 for mitigation effectiveness because of its likely leverage effects.

Sources: Appendix 1 (37); Caldecott *et al.* (2014: 10, projects UR-00458, UR-00481, and UR-00429).

Box 4.9 Renewable energy investments

"The World Bank helped Sri Lanka institute standardised small power purchase agreements that facilitated access to the power grid. Analytical work, capacity building, and demonstration have contributed to Mexican and Chinese adoption of favourable renewable energy payment schemes, which in turn have stimulated more than 20 GW of installed wind capacity in China and hundreds of MW under construction in México. Provision of long-duration loans has a much bigger impact on project bankability than the purchase of carbon credits, at current (i.e. low) carbon prices. As countries increasingly rely on paying price premiums for renewable energy, World Bank and Multilateral Investment Guarantee Agency (MIGA) guarantees against breach of contract and other political risks could be catalytic. Direct WBG investments in wind power have been modest. On average, wind power offers significantly lower economic and carbon returns than hydropower because of high capital costs and often low capacity utilisation. Manufacturing cost reductions at the global level, together with better siting and maintenance, are crucial to increasing the competitiveness of wind and other new renewable energy technologies. The largest single area of off-grid renewable energy investment has been in solar photovoltaics, mostly for home use. These projects can have high economic rates of return but have little impact on GHG reductions because off-grid households use so little energy. At current prices, solar home systems have been successful in a narrow niche market: the off-grid household that is either relatively well-off by rural standards or can access good microfinance services" (Caldecott *et al.*, 2014: 406).

Source: Appendix 1 (45).

Tidal power

The vertical movement of sea-water and the flow of tidal currents through narrow channels offer opportunities to capture some of the ocean's kinetic energy and convert it into electricity. To illustrate the concept, powerful tidal currents known as the Indonesian Throughflow occur in narrow channels between islands in the Nusa Tenggara arc and elsewhere within Indonesia, due to a twice-daily exchange of waters between the Indian and Pacific oceans. The option therefore exists of generating electricity with tidal generators, which might become an attractive investment prospect alongside the installation of an efficient high-voltage direct current transmission grid to supply tidal electricity to major population centres, for instance in Java (Caldecott & Indrawan, 2009, 2010; Caldecott, 2015).

Recycling and other safe waste disposal

Ozone depleting substances (ODS) include the chlorofluorocarbons (CFCs) that are used in the manufacture of insulating foam for refrigerators as well as being the refrigerants themselves. Old refrigerators therefore contain CFCs in their insulation and heat exchangers, which are released when refrigerators are crushed for disposal. Once in the atmosphere, the CFCs erode the ozone layer and act as potent GHGs, with an atmospheric lifetime of several decades. The investments here focus on the safe disposal of ODS, and also the recycling of electronic wastes (e-wastes), which have a total volume that is currently about 50 million tonnes annually and is expected to grow by a third by 2018 (Box 4.10). This waste contains significant amounts of hazardous material (including heavy metals, persistent organic pollutants, and ODS residues), and can cause serious health and environmental impacts if not properly disposed of or recycled.

Box 4.10 Recycling refrigerators and e-wastes in Brazil, India, and South Africa

One project inaugurated the first specialised facility for recycling refrigerators in Brazil, which was given a mitigation effectiveness score of 7 because of the high potency of the CFCs as GHGs. Others supported a series of feasibility studies, in three cities in India, Brazil and South Africa, the aim in each case being to establish a public-access knowledge base on e-waste recycling, analyse the feasibility of sustainable e-waste recycling, and establish a pilot scheme. These were scored 5 for mitigation effectiveness because of their potential, which had not yet been fully realised at the time of assessment.

Sources: Appendix 1 (39); Caldecott *et al.* (2014: 15–16, projects 7F-07029 and UR-00139).

Climate-friendly buildings and building materials

The building sector is a major source of GHG emissions, both directly (through manufacture of construction materials) and indirectly (through design that can increase or reduce the need for heating and cooling systems as a function mainly of insulation). Examples that are typically rated as moderately effective for mitigation purposes include a focus on lower-energy brick-making, the promotion of energy efficiency skills among architects, builders, and regulators, and financial incentives to encourage and enable homeowners to access financing for energy efficiency modernisations of multifamily buildings (Box 4.11).

Box 4.11 Clean building technologies (vertical shaft brick kilns) in Nepal

The project had two components, one focusing on energy efficient brick-kiln technology and the other on 'cost-effective socially and environmentally friendly' (CESEF) building materials. The new brick kilns were 30–40 per cent more energy efficient than traditional ones, produced 70–80 per cent fewer particulate emissions, and the project helped build 26 of them. When the project was assessed, however, most were inoperative or working at less than their full capacity. This was because the new kilns were less profitable and required additional skills to operate as well as higher-quality raw materials. The project tried but failed to relieve these constraints through policy reform. The CESEF component had the potential to reduce emissions by up to 40 per cent, and CESEF technologies were transferred to close to 300 users with varied results. The limited success of the project showed the importance of understanding the business environment before trying to introduce technical changes. Where policy reform is needed to make technological solutions viable, the focus should be on facilitating that, with technology then being used to show how new requirements can be met. The project was rated as moderately effective and given a score of 4.

Sources: Appendix 1 (38); Caldecott *et al.* (2014: 264–267, project 7F-01898).

Improving air quality

Pollution can be reduced by scrubbing or filtering emissions at source, or by increasing efficiency to reduce emissions per unit of production. Emission sources for many air pollutants are the same as those for GHGs, including worn-out, poorly-maintained, obsoletely-designed, or dirtily-fuelled power units, so efficiency investments in particular can also reduce GHG emissions. This is clear in the urban transport and power generation sectors, where improved energy-efficiency technology can reduce particulate, noxious, and GHG emissions simultaneously (Boxes 4.12 and 4.13).

Box 4.12 A clean air programme in Vietnam

The project aimed at mitigating further degradation of air quality in and around Hanoi, by developing an air quality management system through capacity building and institutional strengthening in the areas of policy reform, awareness raising, pilot projects and managing air pollution and emission data. The project was reasonably effective in reducing emissions of particulate matter and GHGs through improved energy efficiency and strengthened emission policies. The work on national and regional policies created a sound basis for future activities targeting emission reduction and prevention in Vietnam, especially in Hanoi. The project also supported awareness-raising campaigns and pilot projects for mitigation (e.g. eco-driving training for truck and taxi drivers), resulting in 15–25 per cent fuel savings in participating companies. Improved energy efficiency at a food processing plant also contributed to mitigation. Increased awareness and understanding of energy efficiency probably had the greatest GHG mitigation effect. The project also developed an emission database for use in planning and monitoring. The process of building this database improved local capacity to monitor emissions, and the project was rated as moderately effective with a score of 4.

Sources: Appendix 1 (38); Caldecott *et al.* (2014: 303–307, project 7F-03833).

Box 4.13 Urban transit systems

"Growing transport demand clogs limited roadway space in the developing world, resulting in severe congestion, air pollution, and GHG emissions. The single largest WBG response has been to support the deployment of bus rapid transit systems, which cost much less than tramways or subways. Key barriers have been the lack of inter-municipality coordination, and opposition by displaced minibus drivers. However, systems have been successfully initiated in Bogotá and Mexico City and are being expanded there and replicated elsewhere (e.g. Lagos, Jakarta). The immediate economic benefits in Mexico City are an estimated 81 per cent economic return and a GHG return of 10 kg/dollar. Larger, sustainable long-term gains will require demand-side management of traffic and rational land use planning" (Caldecott *et al.*, 2014: 407).

Source: Appendix 1 (45).

Heavy equipment rehabilitation and re-use

Items such as trams and trains are large, heavy and expensive in terms of GHG emissions, both to manufacture and to scrap. Despite the GHG emission consequences of relocation, the rehabilitation and re-deployment of such used equipment is an attractive prospect because of subsequent emissions savings (Box 4.14). Improving public transport services is also an important measure to reduce private car use, so has a beneficial effect on urban quality of life, GHG emissions, air quality, etc.

Box 4.14 Used trams in Romania and Ukraine

A portfolio of projects involved rehabilitating and re-deploying used Swiss trams to cities including Vinnitsa in Ukraine and Iasi in Romania, while also providing spare parts, workshop equipment and training for operation and maintenance. These projects were given an overall effectiveness score of 6.

Sources: Appendix 1 (38); Caldecott *et al.* (2014: 12–13, projects UR-00065 in Romania and UR-00305 in Ukraine).

Negative-emission investments

Negative-emission technologies (NETs) exclude GHGs from the air, or remove them from it, and store them out of contact with the biosphere (Caldecott *et al.*, 2015). Options that are being explored tend to focus on the absorption of CO_2:

- *into biomolecules by living plants*, through the fertilisation of marine algae, reforestation (re-planting lost forests), or afforestation (planting new forests);
- *into biomolecules by soil biota* (thus increasing the content of soil carbon, for example through organic farming);
- *into biomolecules and then converted into elemental carbon* with human assistance, by converting plant material into charcoal (or 'biochar') through pyrolysis, which can then be buried; or else

- *by capturing* CO_2 as an exhaust gas from burning fossil fuels or 'biofuels', or from the air, and storing it under pressure in liquid form or as a solid such as a carbonate rock or hydrate (e.g. in geological formations that have previously been emptied of petroleum gases, or on the sea bed).

Aside from reforestation and afforestation, the whole NET approach is in an exploratory phase, awaiting proven techniques that can be subject to economic feasibility and risk assessment, and tested at scale. In principle, however, rather than focusing on CO_2 a more elegant NET strategy would be to find cheap and effective ways to decompose the most important GHGs, CO_2 and methane (CH_4), into their molecular constituents, thus yielding water (H_2O) and oxygen (O_2) for release, and elemental carbon (C) for storage. Attractive aspects of this approach are that elemental carbon is extremely stable when buried under soil or water, and that decomposition and storage could be done in a decentralised way (e.g. at the wind-farm or even the village level). The cheap catalysts that would make the process financially and energetically feasible remain undiscovered, however. Research aimed at finding them would be a rational effort, and could be accelerated by offering an international prize for suitable catalysts that operate at open-air temperatures and pressures, thus broadening participation in the search to include chemists in developing countries, and others with few capital resources. Finding a practicable way to remove billions of tCO_2e quickly from the air is perhaps the most urgently necessary task that humanity has ever faced (Wadhams, 2016). Meanwhile, however, the biochar approach, which also stores elemental carbon, has a number of virtues in terms of community participation and improving the structure, nutrition status and water retention properties of farmland soils. It also offers a way to dispose of substantial amounts of GHGs – for example, NRCT (2014) estimated that for the Congo Republic about 15 per cent of 11.7 million tCO_2e in national emission savings in 2015–2020 could be achieved by converting logging waste into biochar for use on farm soils.

Encouraging investment in low-carbon technologies

Businesses and cleaner production

These investments include the creation of centres of expertise on how to achieve cleaner production in real-life circumstances, typically within factories but also through government advisory roles, in ways that offer cost savings from efficiency or regulatory compliance, and many other benefits from worker, public and environmental health. The global network of national cleaner production centres (NCPCs) developed since 1992 has founding principles and purposes which include seeking direct and indirect GHG emission reductions in the context of a broader sustainable consumption and production (SCP) agenda. There is evidence from UNIDO (2015) that NCPCs are associated with reducing GHG emissions:

- NCPCs and other institutions in nine Asian countries collaborated in a three-year project that demonstrated the application of cleaner production methods for achieving energy savings and GHG reductions in the pulp and paper, cement, iron and steel, chemicals, and ceramic sectors, with GHG emission reductions being verified for 38 demonstration plants at just over one million tCO_2e per year.
- The implementation of several cleaner production options suggested by the NCPC at a small lead foundry in Perú, reduced the lead content in waste by 19 per cent, enabled the recovery of nearly 350 tonnes of lead per year and reduced water and energy consumption, with total GHG emissions reduced by 270 tonnes annually, and investment costs being recovered within several months.
- With the assistance of the NCPC in Sri Lanka, a desiccated coconut mill reduced its waste output by 18 tonnes per year, while achieving considerable reductions in water and energy use, and reducing total GHG emissions by almost 1,000 tCO_2e per year, all due to an investment of less than US$17,000 that yielded annual cost savings of more than US$315,000.
- A paper and tissue products manufacturer in Kenya, with the assistance of the NCPC, implemented a programme to increase waste water recovery and recycling, achieving a 25 per cent reduction in energy consumption, a 50 per cent reduction in water consumption and a 60 per cent reduction in waste water and other wastes, yielding annual savings in excess of US$600,000, with negligible total investment.

The role of an NCPC is to provide a place where companies, government departments, and others, can go to find ideas, guidelines, skills and standards with which to clean up their activities (Box 4.15). Reasons why clients may wish to pay for these services are based on:

- *cost saving* – by taking the NCPC's advice, a company can make savings in energy or raw materials, while also having healthier employees and a better relationship with society;
- *regulation* – laws that require environmental and other standards to be met, with the NCPC being available to advise on how to meet them and to certify compliance with them; and
- *incentives* – such as grants or tax relief targeting cleaner investments.

Thus, if an NCPC is to be sustainable its technical capacity must be developed alongside roles in marketing, supporting government to devise appropriate regulations, and finding ways to incentivise cleaner production. While there is much variation among the NCPCs and in their contexts and challenges, their general influence in favour of mitigation and other benefits seems likely to be significant and increasing.

Box 4.15 A national cleaner production centre (NCPC) in Vietnam

Cleaner production options proposed between 1999 and 2011 by the Vietnam NCPC to 227 companies in six sectors were accepted by most companies and implemented by many of them, resulting in resource savings and financial benefits for companies and a positive impact on the environment. They led to average savings of 7 per cent in electricity, 9 per cent in coal, 7 per cent in fuel oil, 20 per cent in gas, 18 per cent in water and 25 per cent in chemical consumption. Such changes also have multiplier effects, since reducing industrial water use affects the energy costs of pumping, heating and treating water, while well-managed recycling can save a lot of energy, and changing energy mixes (e.g. from coal to methane) can greatly reduce GHG emissions. The project was given a mitigation effectiveness score of 4.

Sources: Appendix 1 (39); Caldecott *et al.* (2014: 280–283, project UZ-00987).

Green credit facilities

These investments include the financing of green credit trust funds (GCTFs) and similar initiatives to stimulate investment in cleaner production technologies (Box 4.16). They tend to synergise with other leverage mechanisms that create access for local producers to investors, markets and global supply chains, or that develop environmental and social risk management (ESRM) guidelines for financial institutions, or that provide technical assistance to encourage financial institutions to invest in energy efficiency and renewable energy (see below).

Box 4.16 A green credit trust fund (GCTF) in Vietnam

A total of 15 Swiss-backed GCTF projects were under way by August 2013, including investments in new and more efficient equipment in the plastics, paper-making and steel recycling sectors, where major savings in the use of energy and water, and in GHG emissions, had been achieved. These GCTF-backed credits can contribute to multiple improvements; for example, new arrangements for recycling scrap steel have helped reduce electricity consumption and the emission of toxic materials such as dioxins, so the project was given a moderate mitigation effectiveness score of 4. This project was part of a portfolio of GCTFs, in Vietnam (2007–2017), Colombia (2003–2005), and Perú (2003–2018), the common purpose of which was to finance investment in cleaner production technologies by guaranteeing 50 per cent of applicable bank loans, and reimbursing a share (up to 25 per cent in Vietnam, 25 per cent or up to US$200,000 in Colombia and Perú) of investment costs depending on previously defined resource consumption and emission indicators. In all cases, the transactions were to be managed by local banks, and the environmental improvements sought and the indicators for them were determined and verified by the NCPC concerned, thus creating a powerful synergy between the NCPC and GCTF interventions. The GCTF in Perú had been longest in implementation and received a higher effectiveness score of 5 following detailed study, suggesting that cumulative impact is to be expected of this kind of investment.

Sources: Appendix 1 (39); Caldecott *et al.* (2014: 284–288, projects under UR-00050).

Environmental and social risk management

Many market fund managers, and the institutions for which they work, have limited awareness and understanding of social and environmental development issues, risks, and associated opportunities for creating shareholder value. This can be corrected by supporting client governments and companies in developing ways to address climate change, that create access for local producers to investors, markets and global supply chains, and/or that introduce market-based solutions to increase access to sustainable infrastructure services, combined with training focused on ESRM guidelines applicable to financial institutions, and building relevant consulting and training capacities (Box 4.17).

Box 4.17 Environmental and social risk management (ESRM) in Asia and Africa

A series of projects that targeted development of ESRM guidelines and training fund managers in their use in Vietnam, Indonesia, China, Thailand and South Africa were considered likely to be effective because of their effects in generating large cumulative investments that adhere to ESRM standards. They were therefore given strong mitigation effectiveness scores of 5.

Sources: Appendix 1 (39); Caldecott *et al.* (2014: 15, 384–386 and 289–293, projects UR-00027, UR-00576, UR-00263 and UR-00593).

Combining cleaner production, green credit and risk management

Small and medium enterprises (SMEs) are often very numerous and collectively drive much (even most) of the economic activity and employment in many countries, but they are much harder to promote cleaner production among than large companies. Progress on engaging with SMEs can nevertheless be made, especially where ESRM, NCPC and GCTF initiatives are all brought together in a coordinated and synergistic way, with provision for outreach at smaller scales, when collectively they can become increasingly effective over time (Box 4.18).

Box 4.18 Synergies among NCPC, GCTF and ESRM investments in Vietnam

When considered separately, these three projects were given moderate effectiveness scores of 4, but all were expected to do better in future. The NCPC project was implemented with UNIDO, and drew on its global experience of NCPC development. The GCTF project was based on prior and similar initiatives in Colombia and Perú, where evaluations had confirmed the soundness and effectiveness of the approach. Unlike these, however, the Vietnam GCTF ran into problems in the 2007–2011 financial crisis, with banks imposing very high interest rates and stringent collateral requirements, despite GCTF guarantees, which inhibited uptake by investors until banking conditions were relaxed in 2012–2013. The Vietnam ESRM project is implemented with IFC, which is managing a regional programme that greatly enhances the leverage of expertise and the exchange of knowledge between countries. Part of this involves a Sustainable Banking Network which includes regulators or industry associations in Vietnam, China, Mongolia, Indonesia, Lao PDR, Thailand, Bangladesh, and the Philippines (as well as Nigeria, Brazil, Colombia, and Perú). Through IFC-facilitated discussions, a model for sustainable emerging-market banking was rapidly being developed, which was expected to amplify further the greening of businesses in Vietnam and elsewhere. This group of projects was assessed as exemplifying the way forward for truly influential and effective national and regional aid portfolios that seek to promote the systematic decarbonisation of the world's economy, and all were given mitigation effectiveness scores of 4 or 5.

Sources: Appendix 1 (39); Caldecott *et al.* (2014: 280–283, 284–288, and 289–293, projects UZ-00987 on NCPC, UR-00050 on GCTF, and UR-00593 on ESRM, all in Vietnam).

Avoiding the stranding of invested assets

Using the connectedness criterion for assessing aid performance focuses attention on the extent to which systems that are being targeted by an investment are connected to other systems that may affect its outcomes but are beyond its control. Another way to put this would be to describe the intervention's vulnerability or resilience to external factors that may, for example, undermine planned development efforts or lead to the 'stranding' (i.e. devaluation) of invested assets. Various climate-related cases might be highlighted in such sectors as agriculture (vulnerable to droughts), infrastructure (vulnerable to floods), health (vulnerable to disease vector distributions), and biodiversity conservation (vulnerable to shifting climatic zones), but one specific example is where fossil fuel investments may decline in value when governments agree to limit the quantity of such fuels that may be burned on climate change grounds.

Although control of fossil fuel reserves is still a key objective of national and corporate policy, and the discovery of new petroleum resources remains a priority of companies in the petroleum sector and their political allies, this thinking is clearly obsolete in light of climate change. This point is starting to cause concern, since the managers of investments might one day be held liable for the stranding of assets that they could and should have predicted (e.g. Carrington, 2015), and is an active area of research (e.g. Smith School, 2015) that is starting to focus on a wider range of stranded asset issues, such as those to do with biodiversity (Caldecott & Jepson, 2014). The notion that entire portfolios of investments may become worthless as a result of climate change is a potent one, which would be expected to encourage investors to think more carefully about the security of their assets and what they can do to safeguard them.

References

Caldecott, B. & Jepson, P. (2014) *Towards a Framework for Protected Area Asset Management: the Project for Protected Area Resilience*. Smith School of Enterprise and the Environment, University of Oxford (Oxford).

Caldecott, B., Lomax, G. & Workman, M. (2015) *Stranded Carbon Assets and Negative Emissions Technologies: Working Paper*. Smith School of Enterprise and the Environment, University of Oxford (Oxford).

Caldecott, J. O. (2015) *Report on Environmental and Ecological Aspects of the Indonesian Seas Large Marine Ecosystem (ISLME) Project*. Food and Agriculture Organization of the United Nations (Rome).

Caldecott, J. O. & Indrawan, M. (2009) *Update of the Country Environmental Profile for Indonesia*. EU Delegation (Jakarta).

Caldecott, J. O. & Indrawan, M. (2010) *Identification and Formulation of EC-Indonesia Climate Change Cooperation*. EU Delegation (Jakarta).

Caldecott, J. O., Dornau, R., Gollan, J., Halonen, M., Saalismaa, N., Schilli, A., Simonett, O., Stuhlberger, C. & Tommila, P. (2014) *Technical Report on Effectiveness of the Swiss International Cooperation in Climate Change Mitigation and Adaptation Interventions 2000–2012*. Mandate of the Swiss Agency for Development and Cooperation (SDC) and the State Secretariat for Economic Affairs (SECO) (Bern).

Carrington, D. (2015) Warning over insurance firms facing big losses over fossil fuels. *The Guardian*, 4 March. www.theguardian.com/environment/2015/mar/03/bank-of-england-warns-of-financial-risk-from-fossil-fuel-investments (consulted 13 Jan 2016).

DNPI & OGJ (2010) *Creating Low Carbon Prosperity in Jambi*. National Council on Climate Change and Office of the Governor of Jambi (Jakarta and Jambi).

NRCT (2014) *Republic of Congo Emission Reduction Program*. Republic of Congo's

National REDD Coordination Team (Brazzaville). www.forestcarbonpartnership.org/republic-congo (consulted 13 Feb 2016).
Smith School (2015) *Stranded Assets Programme*. Smith School of Enterprise and the Environment, University of Oxford (Oxford). www.smithschool.ox.ac.uk/research-programmes/stranded-assets (consulted 14 Feb 2016).
Stafford, K. (2009) *Heart of Borneo: Green Business Network Business Plan 2010–2013*. WWF Indonesia (Jakarta).
Suebu, B. (2009) *A Global Solution: Building a Low Carbon Economy for Papua Province, Indonesia*. Government of Papua (Jayapura).
UNIDO (2015) *Principal Achievements*. United Nations Industrial Development Organization (Vienna). www.unido.org/index.php?id=o5133 (consulted 11 Feb 2016).
Wadhams, P. (2016) *A Farewell to Ice*. Allen Lane (London).
World Bank (2009) *An Evaluation of World Bank Win–Win Energy Policy Reforms Climate Change and the World Bank Group, Phase I*. Independent Evaluation Group, The World Bank (Washington, DC).
World Bank (2010) *Phase II: The Challenge of Low-Carbon Development: Climate Change and the World Bank Group*. Independent Evaluation Group, The World Bank (Washington, DC).

Chapter 5

Ecological mitigation

Tropical forests and REDD+

The road to REDD+

Ideas for linking conservation of tropical forests with global willingness to pay for carbon storage go back at least to the 1980s (Putz & Pinard, 1983). By the early 1990s they were being explored with corporate funders in the power generation sector, which were looking to off-set the GHG emissions of their activities by conserving forests (Moura-Costa, 1996a). Pioneer investors included Applied Energy Systems and the New England Power Company, both US-based, and the Dutch Electricity Generating Board's Forests Absorbing Carbon-dioxide Emissions (FACE) Foundation. Pioneer conservation arrangements with these partners were soon under way in the Malaysian state of Sabah in north Borneo (Marsh, 1992, 1993; Moura-Costa, 1996b), and elsewhere, and within a year or two had become an established part of the conservation planner's toolkit (e.g. through 'Project Biodiversifix' in Guanacaste Conservation Area of Costa Rica – Rodriguez & Gámez, 1994; GCA, 1995; Caldecott & Lovejoy, 1996). It was recognised from the beginning that "forestry will never be a solution for all the environmental problems related to the accumulation of GHGs [but] there are enough areas under forest land or deforested that, if properly managed, could sequester up to 15 per cent of global emissions of carbon dioxide and other GHGs" (Moura-Costa, 1996a: 8). But after the 'Decade of Destruction' of the 1980s (Cowell, 1990), any source of funding for tropical forest conservation was welcome at the time.

The IPCC had drawn attention to the scale of GHG emissions in the forest sector from 1992 onwards, and, helped by lobbying from the conservation community, by the late 1990s this was being discussed at UNFCCC meetings (Box 5.1). A proposal to promote action on tropical deforestation was made at the 2005 Montréal Conference of the Parties (CoP) and received wide support. Other decisions followed, and since the 2010 Cancún CoP, what is now called REDD+ has meant promoting sustainable forest management and enhancing forest carbon stocks (e.g. through natural regeneration, assisted natural regeneration or enrichment planting in natural forests, reducing emissions from deforestation and forest degradation, and conserving carbon stocks), with due attention to co-benefits linked to poverty avoidance, safeguarding indigenous peoples and ecosystem services, and promoting good governance and biodiversity conservation.

Box 5.1 Development of the REDD+ definition

- *Kyoto CoP (1997)*: agreement on protection and enhancement of GHG sinks and reservoirs, sustainable forest management, and forest plantations.
- *Marrakesh CoP (2001)*: agreement that reforestation and afforestation in developing countries can generate carbon credits (under the Kyoto Protocol's Clean Development Mechanism), but reduced deforestation and forest degradation were excluded due to concerns over leakage.
- *Montréal CoP (2005)*: agreement on further study of reducing emissions from deforestation (RED).
- *Bali CoP (2007)*: agreement on the role of conservation, sustainable management of forests and enhancement of forest carbon (REDD+). Launch of the Forest Carbon Partnership Facility (FCPF).
- *Poznań CoP (2008)*: some progress on highlighting the co-benefits of REDD+ to do with poverty, governance, biodiversity, and ecosystem services.
- *Copenhagen CoP (2009)*: some progress on defining and financing REDD.
- *Cancún CoP (2010)*: agreement of the current definition of REDD+, with the inclusion of safeguards and consideration of gender.
- *Warsaw CoP (2013)*: agreement of guidance on modalities and procedures for carbon accounting, monitoring, reporting and verification for REDD+.
- *Paris CoP (2015)*: agreement to put past decisions on REDD+ into effect, to endorse payment by results, joint mitigation and adaptation, and the value of non-carbon benefits, to limit global temperature rise to below 2.0°C, and to commit US$100 billion annually to the climate change response.

Sources: Appendix 1 (49); Caldecott *et al.* (2015).

REDD+ in principle

The idea of REDD+ as a transaction in which companies would pay for forest conservation to off-set their own emissions, conserve biodiversity, and obtain commercial advantages in their home markets, remains valid and useful. Similar approaches have been or are being explored for other ecosystems, such as grasslands, coral reefs, and peatlands, in which avoided degradation can yield measurable emission reductions. Thus inspired, conservationists tried to work out, more or less from first principles but informed by what they had seen in practice, what a REDD system ought to involve if it is likely to be effective as a conservation mechanism (Box 5.2).

Box 5.2 Some principles for REDD transactions

A starting point is that the owners of forest, peatland, karst (limestone) and other carbon-rich areas should earn income by setting those ecosystems and lands beyond use, so that their carbon would never be released, or at least not for many decades. The form of income could vary, but it would need to be related to the amount of carbon stored, shared according to ecosystem ownership, and at least some of it financed by global society. Such benefits would be earned by extending protected area (PA) systems, either permanently or for fixed and prolonged periods. This would create an incentive to fill in gaps in PA coverage, to extend PAs, to complete the legal establishment of proposed PAs, and to create new PAs, in each case through appropriate benefit-sharing partnership with private land owners, local governments, communities, and indigenous peoples with ancestral or traditional claims to the ecosystems concerned.

Various proposals exist for the source of such income, including new international donor funds and the global carbon markets, but regardless of the source, permanent or long-term set-aside would be only one way that this funding could be earned. Another would be the absorption and storage of carbon as a result of re-growing ecosystems after disturbance, once they are protected, although two factors would mean that annual revenues from this source would be small relative to the earning power of set-asides. The first is that carbon is absorbed only slowly compared with the amount of carbon already stored in many intact ecosystems. The second is that carbon removed from or not released into the atmosphere is more valuable sooner rather than later. This is partly because GHGs persist in the atmosphere and exert their influence for decades after release, and partly because the atmosphere-climate system is dynamic and the chief risk is catastrophic destabilisation in the immediate future, rather than gradual change in the distant future. This means that the front-loading of benefits in anticipation of and/or in return for immediate changes in behaviour would be justified, but also that longer-term rewards for continuing results would also be justified. Thus, a substantial share of the total benefit package would have to be paid in advance, the amount and precise pattern of disbursement (i.e. the beneficiaries, the form of transfer involved, etc.) depending on what needs to be done and by whom to set up a system capable of conserving the specified carbon store. In addition, an effective REDD system would need to involve a series of payments conditional on verified land and ecosystem protection and/or verified carbon absorption and storage. Such payments might be spread out over a long period, say of 50 years, and staged in various ways. For PAs, these stages might include:

- *completion of a benefit-sharing contract* among ecosystem owners arrived at through free and informed negotiation;
- *completion of a carbon assessment of the area* and other studies needed for effective management planning (and the development of parallel, non-carbon revenue streams, such as payments for catchment services, ecotourism, and bioprospecting);
- *legal establishment of the area*;
- *completion of an adequate management plan*;
- *establishment and operation of a management service* that is capable of protecting and monitoring the resource; and
- *regular confirmation that the carbon resource is intact* (e.g. every ensuing decade).

The total amount of carbon to be set aside would be calculated at the beginning, and shares allocated to the various resource claimants. The advance payments required to establish the system would be deducted, and the balance would be paid for in tranches representing a share of the carbon valued at prevailing market rates (or in some other way to be agreed) to reward its continued protection.

Protected areas that already existed prior to the REDD arrangement would be assumed to have been established for purposes unrelated to carbon (such as the national patrimony of biodiversity or landscape values), so in principle need not be paid for again with 'carbon money'. However, many existing PAs are under-financed and poorly managed, so a case can be made for increased international investment in PA management, partly justified by avoided carbon emissions, and partly by other global interests. Note also that *all* PAs, whether established with carbon money or not, should seek ways to pay for their management into the future, beyond the time when REDD payments can be justified.

On this basis, a fair and effective REDD mechanism would need to incorporate seven design criteria:

- *Quantity*. The amount of payment must be based on the amount of carbon stored in the ecosystem. The densest carbon stores should attract the highest payments.
- *Duration*. The deal must last for at least 50 years, covering the time when our efforts on emission reduction will be most urgent.
- *Generosity*. Ten thousand hectares of forest might contain 15 million

tonnes of carbon. There will be a big difference in the effectiveness of any deal if the carbon is valued at $5 or $50 per tonne – the difference between $75 million and $750 million spread over 50 years. Thus a high, stable carbon price is vital for the success of this and all other mitigation investments.

- *Immediacy*. The system should be front-loaded to reward carbon storage sooner rather than later, since we need the pressure off GHG emissions now, not later when other systems should be in place to deal with them.
- *Innovation*. If we succeed in fixing the carbon crisis, we will remember that there are many other reasons than carbon to keep forests and other natural ecosystems, including biodiversity, water services, and wilderness values. Conservation will still need to be paid for then, so we should also invest in finding non-carbon revenue sources such as payment for ecosystem services (PES), ecotourism, and bioprospecting.
- *Inclusion*. Most ecosystems are owned, claimed or used by more than one group of people. These will often include central and local government, private land owners, communities, and indigenous peoples. All plausible claimants should be rewarded with a share of the carbon revenues, since otherwise they will not help protect the resource and may even try to destroy it.
- *Verification*. Since one-off, up-front payments in return for promised ecosystem protection will not work, there must be series of conditional payments over the 50 years to reward proven ecosystem protection and/or verified carbon absorption and storage, based on milestones through social agreement, carbon quantification, legal protection, management planning, protective capacity-building, and physical inspection at regular intervals.

Such a system would encourage the extension and completion of PA systems, and if the carbon price were high enough then every surviving patch of forest and peat would end up protected for at least 50 years. These design criteria would ensure that no stakeholder group need be excluded from access to carbon financing, that conditional funding would be available to reward effective conservation, and that the innovative and sustainable use of ecosystems and biodiversity would be encouraged and enabled.

Source: Author.

The idealism and excitement among conservationists in the early years of the REDD debate soon gave way to problems inherent to extending a transactional approach to international relations, in which countries with relatively large amounts of money would pay those with tropical forests to conserve them. This ran into difficulties which could not be entirely overcome by reframing the proposal to stress the multiple co-benefits of wise forest ecosystem management, by arguing that ecosystem goods and services are so valuable to developing countries that being paid to conserve them must be a 'win-win' solution. Some stakeholders rejected the whole idea of a transaction between the weak and the powerful, regardless of benefits to either side. Most others, though, were more interested in how to make it work in their own interests, financially and politically. Hence the original idea did not long survive in simple form, but became entangled with a complex series of negotiations at the international level. But meanwhile it also became clear, to officials and bankers talking with each other about money (if not to conservationists who had known it all along), that conserving tropical forests can be an extremely hard thing to do. Yet it had to be done, convincingly, irreversibly, without causing deforestation somewhere else, and with a realistic calculation of additional GHG emissions saved in the process, or there would be no basis for any kind of REDD+ action, whether described as a transaction or not.

REDD+ in practice

A conclusion from several years of design and trial activity is that a REDD+ process needs to engage with stakeholders that have the capacity to cooperate with one another while fulfilling various complementary roles. They include at least the following groups.

- *Those responsible for conserving carbon on the ground*, which is a complex undertaking that in a rainforest requires secure, mapped and monitored forest boundaries that are actively defended against encroachment, fire and illegal logging. This is often most effectively done by government forestry personnel in alliance with local communities who have an interest in forest protection because they receive financial or other benefits from it.
- *Those responsible for negotiating REDD+ deals and managing money*, each of which requires a particular set of legal and financial skills, organised within distinctive institutional arrangements and backed by appropriate policy and legislation.
- *Those responsible for monitoring, reporting and verifying (MRV) emission reductions*, a function that must be independent of the other roles to avoid conflicts of interest and preserve the credibility of the REDD+ mechanism. The MRV institution would need to monitor and report on status and change in land and forest cover, provide information to the public, access any and all information needed to monitor forest carbon emissions, and build national capacity to measure and monitor activities affecting forest carbon stocks.

Getting all these functions properly established, with all the necessary capacity, technology, policy, and legislative and political support, and working together and aligned to a common end, is hard to do and has not yet been accomplished anywhere. Meanwhile there may be differences among stakeholders and within and between the institutions involved in a REDD+ process on how they see its underlying purpose. These might be papered over, but could manifest later as differences in emphasis, enthusiasm, and cooperation. Opinions may differ, for example, on whose financial, intellectual, or other investments led to which opportunities across the whole REDD+ landscape, which can affect the claims on any potential benefits that may arise. Moreover, stakeholders may diverge in their strategic objectives, especially between obtaining emission reductions as a primary aim with various co-benefits being secondary, and obtaining sustainable livelihoods, biodiversity and ecosystem service benefits as the primary aim, with emission reductions being secondary.

Despite all these challenges, progress has been made where individual forest-owning countries have committed themselves to make the necessary changes to policy, law, and systems of governance and knowledge management. These have had to be focused on what is known as the 'land use, land-use change and forestry' (LULUCF) sector. In this diverse economic cluster, changes to the causes of GHG emissions demand considerable political and other forms of investment by the countries concerned, often against the interests of those who would prefer there to be no change. Finding ways to create the motivation to make these investments is crucial, and two main strategies have been used (plus a third mentioned later). In the first, an opportunity may be detected in a forested country by a potential REDD+ donor, leading them to conclude that the offer of REDD+ financing might be welcomed by the government concerned, an example being the steps that led to the Indonesia–Norway REDD+ Partnership in 2010 (Box 5.3). In the second, a group of like-minded donors may join up to offer forested countries the opportunity to follow a subsidised pathway towards achieving the capacity to make REDD+ arrangements, an example being the Forest Carbon Partnership Facility (Box 5.4). Both are named 'partnerships' to stress that the relationships involved have to be intimate and enabling in both directions if they are to have any effect.

Box 5.3 The Indonesia–Norway REDD+ Partnership

The climate change response. Climate change was propelled into Indonesian public consciousness by the UNFCCC CoP 13 in Bali in December 2007. The message from the meeting was two-fold: that as an archipelagic country, with many people living near coasts or dependent on rain-fed agriculture, Indonesia is extremely vulnerable to sea-level rise and variation in rainfall patterns; and that the country should accept responsibility for its role as a major GHG emitter due to deforestation, peatland degradation, and its increasing reliance on coal-fired energy (Caldecott & Indrawan, 2009). As well as shaping national adaptation planning, these influences on government thinking led to an important milestone in the President's announcement in September 2009 that Indonesia would reduce its GHG emissions by 26 per cent against a business-as-usual estimate of emissions in 2020, and that a further reduction of up to a total of 41 per cent would be made if sufficient international help was forthcoming. It quickly became clear, however, that only by reducing deforestation and peatland degradation could emissions savings be made on a large-enough scale for these targets to be met. Because it was thought that emissions could be reduced at low unit cost by applying REDD+ in Indonesia, the country's forests and peatlands had already attracted both public and private investment, through public funds capitalised by various donors and site-specific REDD+ projects funded by charities, banks, and governments. The Ministry of Forestry had meanwhile asserted its claim that all such projects in the Forest Estate should be managed under its own auspices, and financial benefits distributed at its own discretion. This claim was quickly challenged by civil society and local governments that wished to participate in decisions affecting forest lands, and by the Ministry of Finance which saw REDD+ financing and revenues as being under its own jurisdiction. It also caused unease among expert observers, donors, and investors, since the forestry system was not yet considered fit for the purpose of REDD+ investment, which to work requires credible, verifiable, long-term arrangements whereby known quantities of stored carbon are held securely within newly protected ecosystems. To create suitable conditions would require wholesale reform of the sector, in the face of resistance by groups with strong contrary interests.

The Indonesia–Norway Letter of Intent. High-level political interest in climate change mitigation in Indonesia attracted the attention of Norway's International Climate and Forest Initiative (NICFI). Negotiations then led to the creation of the Indonesia–Norway REDD+ Partnership through the signing of a Letter of Intent by the two governments in May 2010. The aim was to facilitate creation of an Indonesian capacity to deliver quantified and verified carbon mitigation through a REDD+ arrangement that was both marketable and safeguarded. The Letter of Intent committed Norway to contribute up to US$1 billion for this purpose, with the funds to be released in response to results actually achieved, and divided into two parts. The first, of about US$200 million, was to be spent on building the capacity to deliver emission reductions through REDD+, and included an advance commitment of US$30 million to meet start-up costs. The second, of about US$800 million, would be released only if emissions were actually reduced.

The Indonesia–Norway REDD+ Partnership. Three phases of the Partnership were envisioned: Phase 1 for preparation; Phase 2 for system transformation; and Phase 3 for delivering emission reductions. A Joint Concept Note then specified six deliverables for Phase 1, focusing on processes and changes to institutions and laws that were thought necessary before Phases 2 and 3 could begin. The first evaluation of Phase 1 deliverables (Caldecott *et al.*, 2011) found that the whole enterprise was having a strong educational impact, by bringing up many questions that would have to be answered if REDD+ was to be made to work in Indonesia, some of them never having been asked before, and also on the six specific deliverables.

- *Deliverable 1: Establishing a national REDD+ agency.* This would lead the design, oversight, and implementation of a national REDD+ strategy, coordinate all REDD+ initiatives, set up a funding instrument, ensure that conflict and compensation claims are addressed and social and environmental safeguards applied, and build the capacity of local institutions to implement REDD+. The evaluation found that a well-resourced and effective REDD+ Task Force had been set up to prepare the way for the agency, but that progress was slow on overcoming complex technical and, more importantly, political challenges. There were particular difficulties in building consensus on the role and legal mandate of the new agency, in formulating a legal instrument to suspend forest and peatland conversions, and in promoting transparency over maps and data relevant to the Forest Estate, all of which were being contested by established interests.
- *Deliverable 2: Introducing a moratorium on forest and peatland concessions.* The aims here were to reduce damage while other things were done, and to establish a reference point on the extent and condition of forests and peatlands, and lands classified as degraded. The issue of the moratorium was found to have provoked intense media attention and lobbying by civil society and vested interests in the plantation sector, thus delaying the necessary legislation.
- *Deliverable 3: Designing an independent MRV institution.* Establishing the principle of open access to information held by the Ministry of Forestry was proving difficult.
- *Deliverable 4: Establishing a temporary financing mechanism.* Funding would need to be managed according to established international standards that include fiduciary, governance, environmental, and social safeguards, and the mechanism was to operate to the satisfaction of Indonesian authorities. It was found that temporary arrangements had been made for UNDP to advise the Task Force and for McKinsey & Co. to provide facilitation support, while dialogue on a future intermediate-term funding arrangement was under way among the Task Force, the Ministry of Finance, development banks, donor agencies, and civil society organisations.

- *Deliverable 5: Developing a national REDD+ strategy*. The strategy was to be prepared through a credible, transparent, and inclusive process, and should address all the key drivers of GHG emissions from LULUCF, so as to provide clear direction for activities in Phase 2. Although the strategy itself was delayed, it was found that the on-going process of developing it had created new forums where relevant issues were being actively and constructively debated by diverse stakeholders.
- *Deliverable 6: Selecting the first pilot province*. The role of the pilot province was to deliver reduced LULUCF emissions, to set an example for other provinces, to experiment with different REDD+ projects, and to test and refine the newly established institutions. The chosen province, Central Kalimantan in Indonesian Borneo, was found to encapsulate many issues most relevant to REDD+ in Indonesia, including serious problems with spatial planning and numerous existing and often contiguous or overlapping concessions for logging, mining, plantations, and REDD pilot projects. It was also found that building partnerships at the provincial and district levels provided opportunities to make strategic progress on difficult issues that had become blocked at national level.

The second evaluation of Phase 1 deliverables (Caldecott, Rizki *et al.*, 2013) found that there had been progress in establishing arrangements needed for beneficial reforms to governance of the LULUCF sector, and that with the support of the office of the Presidency, other national institutions, sub-national (provincial and district) stakeholders and civil society, and in direct collaboration with UNDP and indirectly with other development partners, the Partnership had been able:

- to seamlessly extend for a further two years an increasingly, though still imperfectly, effective moratorium on new forest concessions for logging and plantations in primary forests;
- to develop national and sub-national REDD+ strategies;
- to facilitate the creation and updating of a publicly accessible digital map of all forests, peatlands and concessions affecting them; and
- to obtain a legal mandate (through Presidential Regulation 62/2013, signed 31 Aug 2013) to create a national REDD+ Management Agency with broad responsibilities and powers to coordinate, synchronise, plan, facilitate, manage, monitor, supervise, and control REDD+ in Indonesia, including to develop with other stakeholders the necessary systems for managing financial and MRV aspects.

Thus, all deliverables foreseen under the first Joint Concept Note (JCN1) of the Partnership had either been accomplished or were in the process of being accomplished by the time of the second evaluation, or were plausibly likely to be so during the Interim Phase. But with an election year in 2014, there was the urgent need to embed irreversibly the new low-carbon institutions and systems within national and sub-national government and society, and to obtain definitive answers to the many questions (e.g. those related to delivery and accounting of GHG emission reductions and equity and effectiveness in incentivising those reductions) that were still hovering around the REDD+ agenda. The second evaluation also observed, however, that "there will be a need very soon to go beyond the concept of doing 'no regret' things (such as enforcing the moratorium, improving the moratorium map and One Map, STRADAs, base-lining emission levels, clarifying and checking the legality of forest licences, etc.) against a rather vague strategic aim, in favour of doing specific things with clear rules in order to achieve a set of specific objectives in the context of a coherent vision. The question is whether that vision is about low-carbon development (LCD), in which case REDD+ is only part of the picture, or is only about REDD+, in which case other issues arise. The decision point is arriving, and whichever fork in the road is taken, there will be many more decision points ahead. These will have profound implications for how the national budget and Partnership funds (and off-setting and LCD investments) will be spent, and what effect they will have, where, when and on whom" (Caldecott, Rizki *et al.*, 2013: 17–18).

Sources: Appendix 1 (32, 36).

Box 5.4 The Forest Carbon Partnership Facility (FCPF)

The FCPF was launched in 2007 at the Bali UNFCCC CoP. It aims to support developing countries in building their capacity to mitigate climate change by reducing GHG emissions from deforestation and forest degradation while adhering to certain environmental and social safeguards (the whole process being known as REDD+ and involving many actions and many stakeholders). To help build REDD+ capacity, the FCPF has two separate but linked funding mechanisms: the Readiness Fund, which finances the efforts of developing countries to prepare for REDD+; and the Carbon Fund, which finances the piloting of specific REDD+ initiatives. In the longer term, a consistent and adequate way will be needed to value and reward emission reductions (or 'carbon conservation', since the main GHGs all contain carbon bonded with other elements), but this is not yet agreed. Meanwhile, the aim for developing countries that have joined the FCPF is to progress from self-funded and/or grant-financed REDD+ preparation, to an arrangement in which they are eligible for payments by results.

The steps involved are mainly 'no-regrets' actions, since they are likely to be useful to a country regardless of any eventual payments, or the precise way in which these payments are calculated or sourced. This is because they involve such measures as clarifying and reforming forest tenure and forest governance, quantifying economic values offered by conserved carbon and co-benefits, understanding the interests of diverse national and global stakeholders, and completing arrangements to manage forests fairly and well for benefits that range from environmental security and ecosystem services to biodiversity, tourism, and bioprospecting. A country signals that it is moving from the readiness support phase to Carbon Fund eligibility by preparing an Emission Reduction Programme Idea Note. This is a crucial milestone since it explains what the country intends to do to deliver quantified emission reductions, based on consultation, study, and planning over several years.

"With 36 [now 47] developing and well-forested countries participating, the FCPF is the most important process in REDD. The FCPF has two separate but complementary funding mechanisms – the Readiness Fund and the Carbon Fund – to achieve its strategic objectives. The FCPF, launched in 2007, has successfully raised in-country awareness, contributed to south-south learning and built capacity and skills on REDD+ issues. The partnership has served strategically in raising the forestry issue onto the UNFCCC agenda as one of the priority issues, and it has served to highlight the socio-economic and environmental interconnections that need to be simultaneously addressed (and the multiple benefits that could be harnessed through REDD+). The FCPF has served to develop and strengthen MRV capacity (including remote sensing approaches) in partner countries and investigated sustainable ways to provide price incentives for forest carbon stock conservation and the sustainable management of forests in developing countries. Evidence from several participating countries such as Perú and Vietnam also highlight achievements on regulation and administrative aspects (including Strategic Environmental and Social Assessment, integrated land-use planning and zoning, improvement of forest tenure security, and enforcement of planning and environmental rules). There are remaining challenges, however, in pricing and carbon markets, including carbon ownership and benefit sharing [but] the FCPF is at the core of international negotiations, and can serve as a pathway for concrete mitigation action. The partnership was therefore given a mitigation effectiveness score of 6" (Caldecott *et al.*, 2014: 49, 391–394, project UR-00372).

Source: Appendix 1 (45).

There is also a 'third way' to progress the REDD+ theme, in which countries may agree to set up a vehicle to support a variety of piloting and capacity-building activities, with the aim of involving numerous stakeholders and trying out various approaches, partly to get people thinking about links between climate change and ecosystem management, and partly to see what works and what can be learned from it. An example of this approach is the Tanzania–Norway Climate Change Partnership (Box 5.5). The resulting portfolio of projects mediated by NGOs illustrates the kind of approach that will be needed world-wide in coming years, i.e. involving tens of thousands of small projects based on local interests and knowledge, adapted to local conditions, and carried out by local people, networked and sharing knowledge with other groups doing similar things. In this case, REDD+ provided an umbrella, but other common themes are possible, from community-based resource management and decentralised governance, to biodiversity and water. In all cases, a key challenge is to find ways for the 'macro' level of global and national aid programming to engage properly with the 'micro' level of community enterprise.

Box 5.5 Pilots of the Tanzania–Norway Climate Change Partnership

- *Participatory forest management.* The Mpingo Conservation & Development Initiative project 'Combining Reduced Emissions from Deforestation and Forest Degradation, Participatory Forest Management and Forest Stewardship Council Certification in South Eastern Tanzania', involved seven communities being helped to establish Village Land Forest Reserves for sustainable management of certified timber.
- *Water catchment forest management.* The African Wildlife Foundation project 'Advancing REDD in the Kolo Hills Forests', emphasised local participation in multiple co-benefits.
- *Inter-village collaboration in forest management.* The Jane Goodall Institute project 'Building REDD Readiness in the Masito Ugalla Ecosystem Pilot Area in Support of Tanzania's National REDD Strategy', involved developing a forest management plan for an area that is both an important catchment for Lake Tanganyika and a key chimpanzee habitat.
- *Sustainable agriculture in dryland areas.* The Tanzania Traditional Energy Development and Environment Organization project 'Community-Based REDD Mechanisms for Sustainable Forest Management in Semi-Arid Areas', promoted rights-based forest regeneration, land use planning, and diverse income generating activities.
- *Protecting forest reserves on village lands.* The joint Tanzania Forest Conservation Group and MJUMITA project 'Making REDD work for

Communities and Forest Conservation in Tanzania', used land-use planning, mapping, advocacy, and Verified Carbon Standard financing to reduce demand on forest resources.

- *Relieving pressure on heavily degraded peri-urban forests.* The Wildlife Conservation Society of Tanzania project 'Piloting REDD in Pugu Kazimzumbwi Forest Reserves', promoted tenure-based livelihoods through boundary demarcation, reforestation, education, and advocacy.
- *Carbon baselining as a contribution to national REDD planning.* The WWF Tanzania Country Office project 'Enhancing Tanzanian Capacity to Deliver Short and Long Term Data on Forest Carbon Stocks across the Country', undertook field-based carbon stock assessments in collaboration with the Tanzania Forestry Service and other national institutions, and trained local people in relevant skills.

Source: Caldecott, Killian *et al.* (2013).

It must be said, finally, that many observers are frustrated by the slow pace of progress inherent to the prevailing strategies of reform, education, and empowerment within the REDD+ agenda, especially in the absence of a convincing way for carbon storage to be paid for adequately, reliably, conditionally, and in the long term. They note that remaining large areas of tropical forests are still being hastily converted to plantations, or fragmented by new roads, and logged or mined, often regardless of national planning or environmental protection laws, because the financial rewards are overwhelmingly attractive to elite groups that have captured key decision-making processes in their own interests. Hence there is a worrying disconnect between the sense of urgency brought on by awareness of climate change and mass extinction, and the glacially slow processes and inadequate means so far deployed to fix the problem.

Forest plantations and farms

Certification of forest plantations and trade in their products

A valid approach to mitigation is to promote the management of tree plantations to credible international standards as defined by the Forest Stewardship Council (FSC), and in the process to increase revenues and benefits flowing to rural people by exploiting price premiums and market access opportunities offered by timber certification (Box 5.6). This should contribute to allowing more durable and equitable management of forest plantations, with some net mitigation benefits.

Box 5.6 Linking trade demand and forest certification

The project aimed to create market linkages between production forests in Vietnam and companies with responsible purchasing policies in Europe, thus encouraging legal and sustainable forest management. By facilitating export-driven compliance with credible international standards of forest management, the project also aimed to provide policy input to the dynamic regulatory frameworks of the forestry sectors of Vietnam, Laos and Cambodia. One of the key components was to facilitate compliance with FSC standards in sustainable forest management for smallholder forests, mainly plantations of the exotic softwood *Acacia mangium*. The project contributed to registration of about 2,000 ha of FSC-certified smallholdings, creating a price premium of up to 43 per cent for *Acacia* wood. A forest carbon accounting exercise by several FSC partners concluded that significant net carbon sequestration was likely within FSC-certified forests over three harvesting cycles (36 years). These calculations suggested an overall mitigation effectiveness score of 4 (moderate), but much would then depend on replication effects, and the extent to which forest stakeholders complied with FSC principles in the long term.

Sources: Appendix 1 (40); Caldecott *et al.* (2014: 18, 294–298, project UR-00015).

Certification of organic farms and trade in their products

Organic farming is known to increase the resilience of agricultural systems and the carbon content of soils, making it attractive to certain donors, and organic products often command marketing premiums that can reward investors, whether private or public (Box 5.7). According to UNEP (2010), in Uganda CO_2 emissions per hectare are up to 68 per cent less on organically than on conventionally farmed land, and organic fields hold 3–8 tonnes more carbon per hectare; there are also co-benefits in terms of biodiversity, water, and human health. A 2009 study by the UK-based Soil Association (2009: 140) found that "organic farming produces an average of around 28 per cent higher soil carbon levels than non-organic farming in northern Europe after around 15 years of organic management", while also improving soil structure and quality, thus supporting climate change adaptation "by reducing the impacts of flooding, droughts, water shortages and desertification". The same source estimated the carbon conservation potential of widespread organic farming to be close to 1.5 billion tonnes (5.5 billion tCO_2e) per year, or about 11 per cent of annual anthropogenic GHG emissions. The diverse benefits of organic farming can make it a useful addition to more conventional agricultural development projects when their impacts on human or ecological health, or GHG emissions, need to be mitigated or off-set (Box 5.8).

Box 5.7 Linking trade demand and organic certification

A series of projects in Burkina Faso, Ukraine, Ethiopia, Ghana, Kyrgyzstan, and Mali aimed to reduce environmental impacts of conventional farming by reducing the use of biocides while improving farmer livelihoods through marketing support, the latter often aimed at promoting the export of organically certified commodities involving domestic or regional buyers, or buyers in European markets for cocoa, cotton, and coffee. The approach variously focused on transferring knowledge about organic techniques and niche market potentials, developing the standards needed to support trademarking and certification, improving the services available locally to organic farmers, and forging relationships between producers and buyers. An additional global finance project provided for the continued availability of trade finance for organic produce from developing countries, by contributing to a guarantee facility in the wake of the 2009 financial crisis (which is a good example of strategic adaptation to changing circumstances). Considering the known mitigation potential of organic farming, and its likely co-benefits in adaptation and human and ecosystem health, these projects were all given effectiveness scores of 6.

Sources: Appendix 1 (40); Caldecott *et al.* (2014: 19, projects UR-00045, UR-00196, UR-00168 for regional exports, UR-00366, UR-00164, UZ-01193 and UR-00152 for European exports, and UR-00419 for global financing continuity).

Box 5.8 Liming soils, GHG emissions, and organic farming in Kenya

The Programme for Agriculture & Livelihoods in Western Communities (PALWECO) supported area-based integrated development in Busia county, near the Uganda border, and its research had established that most people were living below subsistence level on average land holdings of about a third of a hectare. Tsetse fly-borne trypanosomiasis had removed the livestock and sources of dung, and the soils were acidic and nutrient-depleted due to underlying iron rock and long-term farming, with poor cation-exchange capacity that meant that inorganic (NKP) fertilisers were ineffective. The project sought to raise soil pH by administering lime to make the soils more responsive to fertilisers, and thousands of tonnes of pulverised limestone had been distributed to farmers by the time of evaluation. Two main issues with this were identified. First, it would involve significant release of GHGs as lime decomposes in the soil, which would need to be quantified and may justify systematic off-setting by promoting organic farming which is known to increase soil carbon storage. Also, the dispersion of limestone in this way would amount to a major agricultural intervention, which in Kenyan law would be likely to require an environmental impact assessment over and above that required for licensing the limestone quarry itself. Neither observation was hostile to the principle of increasing farmland productivity, but both illustrate the issues that may arise when a broader and more climate-sensitive view is taken of a conventional aid investment.

Sources: Appendix 1 (48); Uusihakala *et al.* (2015: 32).

Using other business incentives

Incentivising changes in business models and behaviour

In 2014, several of Asia's largest agribusiness groups announced new policies on their use of forest ecosystems (Richardson *et al.*, 2014). Wilmar International Ltd renounced deforestation and peatland clearance as parts of its business strategy. Asia Pulp and Paper (APP) committed itself to restoring and protecting a million hectares of rainforest across Indonesia. And Asia Pacific Resources International Ltd (APRIL) announced the development of a 'sustainable forest management policy' with a moratorium on clearing in concessions where there had not been an independent assessment of conservation values. A number of influences had been brought to bear to yield these results (Box 5.9), including the following:

- *Sustained criticism by NGOs of corporations* such as APRIL and APP over their large-scale deforestation activities, and associated social conflicts and human rights violations, leading eventually to constructive dialogue between them.
- *Participation by Wilmar in the Carbon Disclosure Project (CDP).* This helps companies meet the environmental concerns of their stakeholders by researching their supply chains and social policies on behalf of banks and funds that collectively represent many trillions of dollars' worth of assets. The CDP has influence by combining the incentives of shareholder satisfaction, competitive success, and potential investment, with the potential for reputational harm in the event of non-disclosure.
- *The effect of the EU Timber Regulation* (Box 5.10), and the need to convince consumers in Europe and the US that products would not harm biodiversity and climate, coupled with Indonesian government policies and the introduction of a timber legality certification system.

Box 5.9 Consumer preferences in Europe and corporate behaviour in Indonesia

"European consumers have become increasingly conscious of the impact of their consumption habits on global deforestation. European companies such as Unilever, Nestlé and Kingfisher have led the way in transforming their business models to respond to consumer demand for higher standards of environmental protection. In Europe it is now governments who are playing catch-up, as new legislation such as the European Union Timber Regulation seeks to tackle illegal logging, but activists, citizens and politicians are calling for the regulations to go further and to ensure that all timber products entering the EU markets come from a sustainable source. In the last two years, we have seen corporates in the region rise to this challenge with the landmark deforestation commitments made by major companies such as Golden Agri-Resources, Asia Pulp and Paper (APP) and, more recently, Wilmar. In Asia it is the private sector, working with both local and international civil society organisations including Greenpeace, WWF, The Forest Trust and the Rainforest Alliance, that is leading and setting the agenda on what the new standard for sustainability should look like. There is consensus that a meaningful commitment to zero deforestation should be at the heart of this new chapter of responsible corporate behaviour in Indonesia. Of course, there are issues that need addressing. The on-going dialogue over the Indonesian domestic timber licensing system, the SVLK, has shown the capacity gaps in the system that need to be tackled."

Source: Canning (2014: 6).

Box 5.10 The EU Timber Regulation

The EU Timber Regulation (EU 995/2010) requires that due diligence is applied to all timber (including wood products and paper) placed on the EU market, and that traders at all levels of the supply chain keep track of what timber was bought from whom, and in some circumstances who they were sold to. Due diligence comprises: (a) gathering information about the supply of timber, including description, species, country of harvest, quantity, name and address of supplier and trader and documents indicating compliance with the applicable legislation; (b) evaluating the risk of placing illegally harvested timber on the market, using specified criteria (i.e. assurance of compliance with applicable legislation, including certification schemes and third party verification, prevalence of illegal harvesting of specific tree species, prevalence of illegal logging in the country of harvest, UN or EU sanctions on timber imports or exports, and complexity of the supply chain); and (c) unless the risk of illegality is negligible, steps must be taken to mitigate this risk (e.g. gathering additional information, third party verification). Consequences for non-compliance include fines, seizure of the timber and immediate suspension of authorisation to trade.

Source: EC Environment (2016).

The changes just described are a way-station in a long journey that concerns the relationships among Indonesian forest ecosystems and the peoples, businesses, and governments that have sought to use those ecosystems and each other for various purposes. Part of this story is told elsewhere, in relation to the changes of policy that gave rise to the Indonesia–Norway REDD+ Partnership from 2010 (Box 5.3) and its back-story from 1961 (Chapter 8). Another part is the interweaving of various global concerns – for biodiversity and indigenous peoples as well as climate change and sustainable development – with those of Indonesian stakeholders. This is to an extent expressed through long-term forest investments by the UK in a synergistic relationship with the EU's own initiatives, with both players responding at various speeds and in various ways to evolving public interest and opinion. That this aspect has been both very long term and very productive is indicated in Box 5.11, and although it is just a part of the whole story it sheds light on how such strategic undertakings can best be advanced.

Box 5.11 UK tropical forest aid: flexibility, sovereignty, leverage, and patience

"In our opinion the best examples of 'doing the environment' will come when donors put the environment centre stage, engage flexibly with multiple actors on issues that they have prioritized (illegal logging, reducing deforestation, watershed management, etc., in which the poverty–environment linkages are made explicit) and are prepared to adapt as they go. DFID's Forest Governance and Trade programme (and its successor, the Forest Governance, Markets and Climate Programme), is an example of a donor doing the environment well – using the entry point of clamping down on illegal logging through the EU Forest Law Enforcement, Governance and Trade Action Plan – to lever change in broader environmental governance in timber export countries. Reasons why this has worked include:

- the flexible funding to multiple government, private sector, academic and NGO/advocacy groups to ensure multi-stakeholder negotiations;
- the focus on 'legality' supporting national sovereignty and thus palatable to government;

- the use of market leverage (EU restricting trade on illegal products) to bring in private sector;
- linking supply side support with demand side measures;
- allowing money to follow progress not drive process;
- high emphasis on transparency and formal legal status of agreements;
- patient multi-stakeholder processes – with significant funding of gathering and sharing evidence; and
- the attention to careful coordination with other European donors and effective coalitions of interest."

Source: House of Commons (2011: Ev 158–159).

Pricing conserved carbon

Carbon conservation means not only reducing emissions by technological means, and recapturing and storing carbon that has already been released or inevitably will be, but also saving living ecosystems (e.g. forests and soils), and protecting from release the carbon deposited by living ecosystems (e.g. peat and coral), or by long-dead ones (e.g. coal, oil and methane deposits in rock layers and sea beds). All these tasks require investment to pay for protecting deposits and ecosystems, for developing, introducing and incentivising the use of energy sources not based on fossil fuels, and doing the same for carbon capture and storage techniques. The amount of investment needed is astronomical, however – the "multitrillion dollar investments" mentioned by the World Bank (2010) in Chapter 1 – and it is needed urgently. Although governments, charities, and individuals can help, direct investment in conserving carbon at a useful scale is presumed to require the voluntary deployment of private capital. This cannot happen in sufficient quantity at present because there is no predictably high price for conserved carbon assets, and hence no basis for justifying investment in them in terms to which private capital can respond. It seems that markets cannot create such a price themselves, governments cannot agree to establish one collectively, and it is unreasonable to expect enough public pressure on enough governments or companies to make it happen. This conundrum is responsible for the dangerous lack of decisive action on carbon-neutral/negative investment.

If private investment is required to leverage change on an adequate scale then public efforts should target the creation of surety in investment terms, i.e. a long-term, predictable price for avoided or negative carbon emissions, alongside other subsidies in the areas of technology, governance and the deliberate de-risking of investments through underwriting mechanisms. In short, game-changing alterations in the investment context are essential if private funds are to catalyse low-carbon development through REDD+, or any other means. So the central issue is not how public investment can reward REDD+, but rather how public investment can create the conditions needed for private investment to reward and profit from low-carbon development, including through REDD+. Initiatives such as the Indonesia–Norway REDD+ Partnership and the FCPF focus on creating the means by which countries can conserve carbon, justified partly by the 'no-regrets' benefits of doing so, and partly on the assumption that market conditions will soon change to allow for emissions savings to be paid for.

It is self-evident, however, that if the true cost of GHG emissions were factored into the price of energy and all items made or grown using it, then the price of these things to their users would rise. But there seems to be little political appetite either to impose a realistic pricing structure or to ensure that, having done so, poor people obtain adequate supplies. And yet, something of the sort is clearly needed, since it is impossible either to negotiate with nature or to ignore for ever the consequences of mass deprivation. The demands of people, currently expressed through forms of activism illustrated earlier in this chapter, but soon likely to intensify and diversify, must be met if remotely possible. A redefinition of what is meant by 'possible' is what all this is actually about, and business as usual (i.e. a barely regulated capitalist system that degrades ecosystems and relishes inequality) is clearly on the wrong side of the debate. For a public-sector development actor, the implication is that investments should target three areas:

- *promoting public understanding* of the issues, threats, opportunities, and options, through universal knowledge sharing and the free flow of information and analysis;
- *developing, testing, and improving technological and ecological solutions*, and ways to overcome organisational, fiscal, and other barriers to their adoption; and
- *exploring new ways to transform economic systems* in favour of realistic energy pricing and equity, while overcoming and out-manoeuvring political barriers to change.

The last part is the most difficult and most necessary, but least amenable to demonstration. We know only that there are absolute boundaries, of which the most important are that tens of billions of tonnes of tCO_2e must be removed in net terms from the atmosphere each year for the foreseeable future, that the integrity of ecosystems must be preserved and restored at landscape and continental scales, and that human systems must be made and kept robust and resilient enough to absorb extreme, diverse, and unpredictable environmental and political shocks. We also know that *a price of any magnitude is worth paying for these outcomes*, but that there is currently a *low overall willingness to pay any price at all for them.* In these circumstances the evaluation criteria applicable to any investment would need to value measurable effectiveness at scale, easy replicability, and strong impact and sustainability through rapid and irreversible change in incentives, rules, and relationships that affect collective human behaviour. A thought experiment for an investment that could score highly in these ways is summarised in Box 5.12.

Box 5.12 Creating a useful carbon price

It has proved difficult to ensure a price for conserved carbon that is high and stable enough to facilitate decarbonisation investment. Markets cannot do this alone because speculation undermines predictability, and a universal inter-governmental agreement has proved elusive. A way forward might lie in deliberate price creation by a small number of like-minded governments, involving them in announcing their intention to buy, at a fixed date in the distant future, a large amount of certified conserved carbon at a price high enough to promote rapid worldwide decarbonisation. The amount, date and price would all need to be carefully considered against the criterion of what is likely to work in terms of quickly halting and reversing net carbon emissions (with the year 2060, 500 GtCO_2e, and US$100 as candidates). If the participating governments had sufficient resources for this pledge to be plausible, it should set a long-term floor price for the duration of the commitment, starting immediately. The liability would also need to be considered, but bearing in mind: (a) that the market price of conserved carbon may reach or exceed the floor price by the date set; (b) that the governments concerned could save up to meet the liability by investing in decarbonisation assets that would be expected to appreciate in value; and (c) *if we have not solved the climate change problem in a few decades, nothing will be of value anyway*. The same outcome could be achieved if a charitable foundation obtained enough individual pledges of the same kind, which would be slower and less certain, but desirable since it would share participation in solving the climate problem, in the same way that everyone is responsible for creating it.

Source: Author.

Investing in protected areas

The economic value of protected areas has long been appreciated at a public-policy level, justifying for example World Bank investment in them (Box 5.13). Such sites are also seen increasingly as attractive for long-term diversified investment in ecosystem goods and services (carbon, biodiversity, bioprospecting, ecotourism, water, etc.), especially in places where national sovereignty, a consensus wish to achieve prosperity by using ecological assets sustainably, and extremely high biodiversity and valuable ecosystem services all coincide. This combination of conditions is not common at present, but cases where supportive policies are in place include Costa Rica (Chapter 1), with other countries and territories (e.g. Bhutan, Fiji, Ghana, Guyana, New Zealand, Timor Leste, Sabah) seeming to be exploring the options for sustainable development based on wise use of their own ecosystems. Countries are complex systems, however, and these investment opportunities are inherently founded on local ecosystems and local societies, which are unique and must be understood in detail if they are to be appreciated, much less profited from.

Box 5.13 Forest protection investments

"Forest loss, especially in the tropics, generates a quarter of developing countries' emissions. The local and global values of standing forests often greatly exceed the gains from destroying those forests, and tapping this value can offer large economic and GHG gains. ... World Bank experience provides some models for scaling up; for example, payments for ecosystem services (PES) arrangements also seek to reward property owners who maintain forests, while World Bank-supported programmes in Costa Rica and México have demonstrated the logistics of PES arrangements and have helped to popularise this approach globally. Yet many of the payments have gone to areas that are not at high risk of deforestation, diluting carbon and environmental benefits and prompting attention to targeting. The most prominent line of action associated with forest conservation is support for protected areas. These now cover more than a quarter of the tropical forest estate, much of it with World Bank support. A global analysis shows that these areas are on average effective in reducing deforestation, that areas allowing sustainable use are more effective than strictly protected areas, and that areas controlled by people indigenous to them are the most effective of all, while they all also offer biodiversity co-benefits. These findings support the feasibility of the REDD+ agenda in combining sustainable development and forest conservation" (Caldecott *et al.*, 2014: 407).

Source: Appendix 1 (45).

Protected areas offer a similar investment prospect to that of an ecologically oriented country, since they are a type of ecological asset which by virtue of law and public support have a degree of autonomy and permanence (Caldecott & Jepson, 2014). But, as has become clear when considering REDD+, for investment purposes it is vital to be able to guarantee both a high level of *asset security* (based on legal protection and political support maintained by incentives and education that effectively disable all threats to the reserve), and high levels of *potential value* (catchment values, species richness, endemism rates, amenity, carbon, etc.). Having identified the protected areas that meet these requirements, business plans can be developed for each of their asset dimensions in the context of available technology and current and anticipated market opportunities, and the benefit-sharing arrangements that will be needed for long-term success (Boxes 5.14 and 5.15). In doing so, complex issues of resource ownership and intellectual property rights must be addressed if equitable partnerships with local stakeholders are to work to everyone's advantage (Box 5.16).

Box 5.14 Business planning for the Iwokrama rainforest in Guyana

The Iwokrama initiative comprises an institution (the Iwokrama International Centre for Rain Forest Conservation and Development) and 360,000 hectares of tropical moist forest (the Iwokrama rainforest). It has its origin in the 1989 Commonwealth Heads of Government meeting, at which the Government of Guyana offered the international community a proving ground for techniques of conservation and sustainable use. Funding was made available by the UK, Canada, the EC, and International Tropical Timber Organization (ITTO).

"The Iwokrama rainforest grows on poor, ancient soils, and is both fragile and unproductive when viewed as a resource for industrial logging or plantation agriculture. Viewed from the perspective of the sunrise industries of bioprospecting and ecotourism, however, then the forest has strong comparative advantage as an object of investment. This is because technology and markets have evolved to the point at which they can value and use the chief attributes of the area, which are high biodiversity and natural splendour. The world's economy is becoming daily more diverse, and hungers for yet more diversity, driven by technology that can handle unprecedented amounts of information. The Iwokrama rainforest is an evolved system that contains an extremely dense accumulation of information. Thus it is expected that human ingenuity, modern technology, and the high biodiversity of the Iwokrama forest, put together, can support a spectacular diversification of the economy. Iwokrama's strategy, therefore, is to preserve the biotic resource while making it available for non-destructive commercial use. The expectation is that an expanding array of specialised businesses can and will develop to make creative use of the forest and its information content. The key challenge here is to establish conditions that allow private companies to do what they do best – to take risks creatively for profit – while minimizing conflict and resource degradation, and ensuring that a share of created wealth is retained to finance both the conservation process and the local human development process" (Caldecott *et al.*, 2001: 61–62). The authors of the Iwokrama Business Plan envisioned that revenues would be captured from commercial activities using forest resources, and they analysed opportunities linked to timber and reduced-impact logging, ecotourism, non-timber forest products, bioprospecting, carbon sequestration and trading, and endowments and grants, the aim being a substantial degree of self-sufficiency in core funding.

Source: Appendix 1 (1).

Box 5.15 Business planning for the Maliau Basin Conservation Area in Sabah

The 588 sq. km Maliau Basin Conservation Area (MBCA) in Sabah (Malaysian north Borneo) is named for the Maliau Basin, a nearly circular structure, up to 25 km in diameter and more than 1,000 m deep, bounded by a rim at about 1,600 m above sea level and connected at low altitude with the outside world only through the Maliau Gorge. The conservation area includes the basin itself and its outer slopes, and is one of the last major tracts of pristine rain forest in Borneo. There are 10–12 vegetation types and a complex system of rivers with what is probably the world's densest array of waterfalls. It is located, biologically speaking, in the richest part of the richest tropical island in the world, and is estimated to include about 38 per cent of Borneo's entire biota. The following investment opportunities were identified:

- *Ecotourism*, involving charging local and foreign visitors for access and services within the MBCA, and managing the resulting impacts to minimise conflicts with conservation priorities.
- *Educational services*, for example the sale of subscriptions to web-cast or satellite-broadcast lectures about the rain forest, aimed at the international school and university markets.
- *Educational merchandising*, for example the sale of products such as books, booklets, magazines, postcards, video CDs and posters about the rain forest through international catalogue orders.
- *Bioprospecting*, involving the development of long-term equitable partnerships with groups that undertake commercial research on biodiversity in the fields of biotechnology, pharmaceuticals, cosmetics, etc.
- *Biodiversity futures trading*, involving the sale of rights to use the MBCA's biodiversity resources sustainably in the distant future, to investors who expect the value of those rights to rise in the near future (as biodiversity resources decline with mass extinction), with price increments being taxed.
- *Carbon storage*, involving obtaining international grants or carbon emission credits for re-planting native trees in the buffer zone, or for putting the Maliau coal deposits and forests 'beyond use' under legal protection.
- *Grants, sponsorships, and partnerships*, involving a systematic quest for international grant financing and the recruitment of corporate sponsors and partners for discrete investments, such as educational, library, computing and research facilities, bridges, tree towers and canopy walkways, for long-term relationship building, and for senior staff secondments.

- *Use of trust funds*, involving managing endowments to finance all aspects of the long-term management of the MBCA.

Source: Caldecott (2002b).

Box 5.16 Stakeholder partnerships in the Iwokrama rainforest in Guyana

The Iwokrama rainforest is inhabited and used by rights holders that "have always been recognised by Iwokrama as comprising members of Amerindian communities in the North Rupununi, and in the last year recognition has effectively been extended to a number of groups in Region VIII. These rights holders thus represent the Amerindian half of the Iwokrama–Amerindian partnership. This partnership is deeper and broader than any written agreement can capture, since it is based on a whole array of personal relationships, areas of intra- and inter-group consensus, common interest, trust and the expectation of mutual support in adversity and opportunity. Nevertheless, as Iwokrama moves to establish commercial ventures involving use of the forest's resources, it is necessary for the partners to become much more specific about their respective identities, roles, rights and responsibilities" (Caldecott, 2002a: 8).

Hence sustainable development in the Iwokrama rainforest involves recognising the importance of "the terms on which Iwokrama obtains information from nature and from people's minds, extracts materials from the forest, and gains or authorizes access to the forest. These terms cannot be decided unilaterally by any one individual or institution, since in all cases multiple rights-holders are involved. Instead they are dictated partly by prevailing national law, partly by Iwokrama's wish to abide by international law and best practices, and partly by negotiation and contractual agreement between Iwokrama and its partners. The terms in each case are expressed in the agreements (i.e. contracts) that Iwokrama has with its collaborators, which comprise other institutions, communities and individuals. The protocols of these agreements are the various points to which the parties bind themselves in each case" (Caldecott, 2002a: 3).

Source: Appendix 1 (1).

References

Caldecott, B. & Jepson, P. (2014) *Towards a Framework for Protected Area Asset Management: the Project for Protected Area Resilience*. Smith School of Enterprise and the Environment, University of Oxford (Oxford).

Caldecott, J. O. (2002a) *Partnerships and Sustainable Income Generation at Iwokrama: a Report on Intellectual Property Rights, Access and Benefit Sharing with Particular Reference to Bioprospecting.* Iwokrama International Centre for Rain Forest Conservation and Development (Georgetown).

Caldecott, J. O. (2002b) *Strategic Plan for the Management of Maliau Basin Conservation Area. Management of Maliau Basin Conservation Area Project.* Sabah Foundation (Kota Kinabalu).

Caldecott, J. O. & Indrawan, M. (2009) *Update of the Country Environmental Profile for Indonesia.* EU Delegation (Jakarta).

Caldecott, J. O. & Lovejoy, A. (1996) Costa Rica. Pages 18–27 in *Decentralization and Biodiversity Conservation* (edited by E. Lutz & J. Caldecott). The World Bank (Washington, DC).

Caldecott, J. O., Kowal, T. M., Holland, J. D. & Moran, D. (2001). *Report of the Multi-donor Mid-term Review of the Iwokrama International Centre for Rain Forest Conservation and Development, Guyana.* Department for International Development (London).

Caldecott, J. O., Indrawan, M., Rinne, P. & Halonen, M. (2011) *Indonesia–Norway REDD+ Partnership: First Evaluation of Deliverables.* Government of Norway's International Climate and Forest Initiative (Oslo).

Caldecott, J. O., Killian, B., Tommila, P., Rinne, P., Halonen, M. & Oja, L. (2013) *Scoping Mission for a Possible Renewable Natural Resource Economic Governance Programme in Tanzania: Final Report.* Norwegian Agency for Development Cooperation (Oslo).

Caldecott, J. O., Rizki, D., Rinne, P. & Halonen, M. (2013) *Indonesia–Norway REDD+ Partnership: Second Evaluation of Deliverables.* Government of Norway's International Climate and Forest Initiative (Oslo).

Caldecott, J. O., Dornau, R., Gollan, J., Halonen, M., Saalismaa, N., Schilli, A., Simonett, O., Stuhlberger, C. & Tommila, P. (2014) *Technical Report on Effectiveness of the Swiss International Cooperation in Climate Change Mitigation and Adaptation Interventions 2000–2012.* Mandate of the Swiss Agency for Development and Cooperation (SDC) and the State Secretariat for Economic Affairs (SECO) (Bern).

Caldecott, J., Robledo, C. & Clarke, M. (2015) *The World Bank: Second Program Evaluation of the Forest Carbon Partnership Facility (FCPF), Inception Report.* Indufor Oy (Helsinki) and Forest Carbon Partnership Facility (Washington, DC). www.forestcarbonpartnership.org/sites/fcp/files/2015/November/FCPF%20Inception%20report%20Final%20Nov%202015.pdf (consulted 26 Mar 2016).

Canning, M. (2014) Companies See the Wood for the Trees: Observations by the UK Ambassador to Indonesia. *The Jakarta Post,* 19 May.

Cowell, A. (1990) *The Decade of Destruction: the Crusade to Save the Amazon Rain Forest.* H. Holt (New York).

EC Environment (2016) *Timber Regulation.* ec.europa.eu/environment/forests/timber_regulation.htm (consulted 12 Sep 2016).

GCA (1995) *Project BIODIVERSIFIX: WETFIX and DRYFIX.* Guanacaste Conservation Area (Liberia, Costa Rica).

House of Commons (2011) *The Impact of UK Overseas Aid on Environmental Protection and Climate Change Adaptation and Mitigation: Fifth Report of Session 2010–12, Volume 1.* Environmental Audit Committee. The Stationery Office (London).

Marsh, C. W. (1992) CO_2 offsets as a potential funding source for improved tropical forest management. *Tropical Forest Management Update* **5**(2): 15 and **7**(2): 2.

Marsh, C. W. (1993) Carbon dioxide offsets as potential funding for improved tropical

forest management. *Oryx* **27**(1): 2–3.
Moura Costa, P. (1996a) Tropical Forestry Practices For Carbon Sequestration. Pages 308–334 in *Dipterocarp Forest EcoSystems – Towards Sustainable Management* (edited by A. Schulte & D. Schone). World Scientific (Singapore).
Moura Costa, P. (1996b) Tropical forestry practices for carbon sequestration – a review and a case study from Southeast Asia. *Ambio* **25**: 279–283.
Putz, F. E. & Pinard, M. A. (1983) Reduced impact logging as carbon-offset method. *Conservation Biology* **7**(4): 755–757.
Richardson, L., Caldecott, J. O. & Hakim, S. (2014) *Final Report of the Evaluation of DFID's Forest Governance, Markets and Climate Programme in Indonesia.* ITAD & Triple Line (London).
Rodriguez, C. M. & Gámez, R. (1994) *Science-Society Integration in the 1994 'pre-planning phase' for an All Taxa Biodiversity Inventory (ATBI) of the Guanacaste Conservation Area in Costa Rica.* National ATBI Planning Commission and National Biodiversity Institute (Heredia, Costa Rica).
Soil Association (2009) *Soil Carbon and Organic Farming.* The Soil Association (Bristol). www.soilassociation.org/LinkClick.aspx?fileticket=SSnOCMoqrXs%3D&tabid=132 (consulted 14 Feb 2016).
UNEP (2010) *Green Economy: Developing Countries Success Stories* (UNEP, Nairobi). www.unep.org/pdf/greeneconomy_successstories.pdf (consulted 14 Feb 2016).
Uusihakala, J., Caldecott, J., Killian, B. & Njagi, J. W. (2015) *Finland's Development Cooperation With Kenya in 2007–2013.* Evaluation report 2015: 2. Ministry for Foreign Affairs (Helsinki).
World Bank (2010) *Phase II: The Challenge of Low-Carbon Development: Climate Change and the World Bank Group.* Independent Evaluation Group, The World Bank (Washington, DC).

Chapter 6

Adaptation by preserving resilience and diversity

Natural ecosystems

Overview

Natural ecosystems are those that are not grossly disturbed by people and that retain most of their native species, gene pools, ecological relationships, and intrinsic evolutionary processes. Many of their roles are not universally appreciated, particularly in sustaining biodiversity and providing services as water catchments and carbon stores. Unappreciated public goods tend to be neglected and lost, so building public interest in natural ecosystems is necessary. It is done through environmental education on the value of ecosystem goods and services, and by making ownership and benefit-sharing arrangements involving law, policy, and commerce that reward sustainable behaviours. Such investments help to maintain the resilience of living systems in the face of damage, and the continued availability of those systems as resources to use in adapting to the consequences of that damage, including climate change. The following sections explain ways to safeguard natural ecosystems as key parts of any climate change adaptation process or policy.

Sustainable forest management

Tropical forests can be managed for harvests of timber and many other outputs, ranging from rattan canes, wild meat and biodiversity information, to ecological and economic services. These can be obtained more or less destructively, with a greater or lesser degree of competition with one another, and with more or less attention to future harvests as compared to current ones. Thus forest management involves choices. Conventional industrial logging is focused on a single, maximal harvest of the most valuable timber species to the exclusion of all other factors, and can therefore be very destructive. A commitment to 'sustainable forest management', however, implies that choices are made with a view to obtaining multiple outputs over extended periods. At the least, this involves using reduced-impact logging techniques to minimise damage during the logging process, so that a greater degree of forest recovery is possible, and economic and ecological costs are reduced. At the most, it combines the most careful, precision-harvesting of timber and all other resources for which markets exist or are anticipated, all guided by deep knowledge of the ecosystem being managed.

Many factors influence forest managers in choosing where they want to be on this spectrum, not least the wishes of regulators and the expectations of shareholders. Careful forest management aiming for multiple outputs with a long-term perspective is more expensive than other systems in the short term, so a critical issue is who bears the costs and who receives the benefits. In practice, it is the company with a timber concession that is expected to reduce its short-term profits in the interests of sustainability, while it is society as a whole that would receive the economic benefits if it did so. This becomes an impossible equation if there is either weakness in the regulatory environment or a lack of permanence in the tenure of the concessionaire. For these reasons, while numerous studies have shown how in principle to move forest management towards sustainability, it has been very hard to make it actually do so in practice.

One way forward, which is being tried in Indonesia and elsewhere, is to establish large, permanent forest management units (FMUs) which are to be managed by concessionaires with long-term leases (Caldecott & Indrawan, 2009). The idea is that the holder of each concession would have an incentive to plan for long-term rather than just short-term profitability. Each FMU requires detailed assessment of the area's complete potential to support FMU-specific management planning, including timber resources, biodiversity, ecosystem services, cultural priorities, ecotourism potential, and carbon storage. The idea of basing the management of forest areas not on maximum timber yield, sustainable or not, but on the design of a stand-specific mix of diverse harvests and outputs based on what that particular stand is good for, with detailed research preceding decision making, has long been advocated by foresters and conservationists (e.g. Caldecott, 1988).

Changing forest values: climate, microclimate, and fire

Calculations by forest concessionaires interested in profits, and by governments interested in timber royalties, have tended to be guided by market conditions derived conservatively from past practices. In the 1950s, when professional foresters had most influence in setting up government forest departments in the tropics, and the 1960s to 1980s, when industrial logging peaked, forests were seen as stands of timber awaiting conversion into logs, sawn wood, plywood, pulp, paper, and furniture (Ross, 2001). These habits of thought persisted, but 'followed the money' in new ways in the 1990s to 2010s, when it often became more profitable to convert logged forests to plantations, rather than waiting for them to re-grow timber. For much of this 60-year period, factors often missing from the thoughts of those managing tropical forests included the rights and fates of forest-dependent people and non-human organisms, the economic and potential financial values of non-timber forest goods and services, and the likely consequences of climate change including the vulnerability of rainforests to drought and fire. All these factors are now more prominent, and there may be opportunities to manage surviving natural forests in ways that reconcile new priorities. For example, a company with a very long-term forest concession, or a community with recognised forest rights, may want to obtain money from a variety of forest uses including but not limited to careful logging.

Here the forest owners may wish to consider the need to preserve a fire-resistant microclimate, so as to safeguard the forest itself and its catchment services, or to qualify for carbon storage benefits, and viable populations of large numbers of native species, to allow for ecotourism and bioprospecting, or both. This requires attention to ecology. The importance of preserving moist microclimates in rainforests is underscored by two realities. First, as shown across Indonesia and in Tasmania during the 2015–2016 drought (and in previous droughts, in 1982–1983, 1997–1998, 2002–2003, 2006–2007, 2010–2011, all of them driven by the El Niño-Southern Oscillation in Pacific Ocean currents and winds), ever-wet forests will dry out and become vulnerable to fire when their interiors are exposed to warm, dry, moving air. Intact rainforests are relatively immune to this, except around the sites of tree-falls and land-slips, which are small and can be recolonised quickly from their edges as the microclimate recovers through regeneration. Many of the species present are therefore fully adapted to ever-wet conditions, and the organisms belonging to them cannot survive drying out. Where a tropical forest is naturally exposed to occasional hurricanes (as in the Atlantic forests of Nicaragua), or to regular fires (as in much of Australia), species show characteristic adaptations and can quickly seed or regrow after disturbance. Similarly, very cold air descending occasionally from the Andes into parts of the Peruvian Amazon (e.g. at Tambopata in Madre de Dios) does not kill the local species but induces torpor or other survival responses.

Traditional, low-frequency slash-and-burn farming mimics natural tree falls in rain forests, with small temporary patch sizes and rapid re-growth and colonisation. In this way, species richness is maintained over large areas of forest in a dynamic mosaic of undisturbed and old and new regenerating patches. But this is not the case with mechanised logging at a landscape scale, in which large areas are fragmented by roads and skid trails, most of the closed canopy is broken open, and only few, small patches are left undamaged. Here the hot, dry, moving air outside the canopy can penetrate to ground level, killing fragile organisms, and no moist sites remain nearby from which those locally extinct species can recolonise. And once an ever-wet forest is broken open and fragmented, not only will it lose its fragile species, but fire can easily enter from nearby cleared land to destroy the rest of the ecosystem as well.

In some cases and circumstances it may be possible to preserve large-enough patches of forest microclimate, well-enough distributed, to maintain both the full species richness of a landscape and its resistance to fire. This can be critical if a REDD+, bioprospecting, ecotourism or other arrangement is foreseen as a source of income to the forest owners, or if the forest is sacred or vital for catchment protection or other purposes. A little careful logging might still be possible, but only in the context of a management strategy that takes proper care to safeguard the continuity of forest microclimates. Box 6.1 summarises a simple way to identify forest patches that are large enough to sustain an ever-wet microclimate, which might be done: for planning (e.g. to set aside biodiversity and microclimate refuges); to demonstrate the integrity of a forest's microclimate should it be useful to do so for valuation (e.g. in establishing eligibility for certain payments for ecosystem services, including REDD+); and/or to monitor changes that might occur (e.g. to prompt increased fire defences when necessary). The findings suggest that a patch of several hundred hectares is needed to preserve an ever-wet microclimate in a rainforest, and to act as a recolonisation source for fragile organisms in surrounding areas.

Box 6.1 Microclimate bioindicators and biodiversity reservoirs in rainforests

Many rainforest organisms cannot survive outside a stable, moist, and cool microclimate under the protection of an intact forest canopy, including many species of liverworts, filmy ferns, mosses, flatworms, land-leeches, amphibians, and crustaceans. Hot, dry microclimates can invade virgin forests from their edges, implying that there is a minimum size for a biodiversity reservoir if it is to be effective, and also that patches should be shaped so as to have a low edge to area ratio. Some organisms are water-permeable (i.e. vulnerable to drying out), macroscopic (i.e. visible to the naked eye), and terrestrial (i.e. found near the forest floor), or WaPMaTs, so they are convenient bioindicators of prevailing microclimates. In a study in the Malayan Peninsula, 100 trees at each of several sites were examined for mosses and liverworts growing on the bole between the ground and eye level. Their abundance was scored between '0' (absent) and '3' (abundant), and qualitative observations were also made on the general nature of the forest and its visible biota, with particular attention to leeches. The study confirmed that mosses, liverworts and leeches were common in a large area of unlogged forest and rare or absent in recently logged forest and in isolated patches of virgin forest less than 100 ha in area. These findings were made quickly by Forest Department staff with virtually no preparation, training or special equipment. Implications include that there is a minimum patch size (probably of 200–500 ha) needed to retain a virgin forest microclimate, and that edge effects within patches and recolonisation effects outside them are significant and measurable using WaPMaT bioindicators. To preserve species richness and a degree of fire resistance in a forest landscape, intact patches scattered through the area would be needed, each of a minimum size and compact shape, and dispersed to promote recolonisation of surrounding disturbed forests. Such arrangements would be necessary to allow forest management to be certified as biodiversity friendly, and for forests to resist reversal risks from fire in REDD+ and other payment for ecosystem services (PES) arrangements.

Source: Caldecott (1997).

Multi-stakeholder forest management

Some investments aim to involve, educate and reinforce through policy and law the role of local people in forest ecosystem management. The entry point might be an emphasis on biodiversity, agrobiodiversity, protected areas and their buffer zones, water catchments, coastal reforestation, community-based forestry, and/or national forestry systems that have a social forestry dimension (Boxes 6.2 and 6.3). Such projects are typically systemic in their approach, often reinforcing local actions by raising awareness among local and national authorities, and private stakeholders, of the value of forest ecosystems, and increasing their capacity to conserve them by supporting the design of appropriate policies, regulations and instruments.

Box 6.2 Community forestry in Nepal

The project encouraged and enabled community groups in four of Nepal's poorest districts to adopt sustainable forestry practices, promoting forestry-related employment and entrepreneurship, involving the poorest households in community forestry, and creating a model for village governance through which best practices could be promoted. Consistent with the beneficial effects of other community-based forestry management actions in Nepal, the project led to measured increments in new forest area (of nearly 33 per cent) and forest quality (of 20 per cent), and it was scored 5 for adaptation and mitigation.

Sources: Appendix 1 (40); Caldecott *et al.* (2014: 16–17, 249–254, project 7F-03128).

Box 6.3 Sustainable development at landscape level

A series of projects took a systemic and people-centred approach to area-based sustainable development, but with an emphasis on applied biodiversity research and conservation. They involved multiple stakeholders in strengthening strategy, planning and implementation capacity, ecological gap analyses and sensitivity maps, coupled with increased public awareness and the promotion of positive economic links between conservation areas and society. All the multi-stakeholder forest management projects have so many 'moving parts' (ecosystems, individuals, communities, businesses, governmental institutions, NGOs, etc.), any and all of which can influence outcomes, that it is not possible to score their effectiveness definitively without more comprehensive study in each case. The soundness of the overall approach was, however, recognised with a moderate effectiveness score of 4, noting that forest conservation outcomes have both mitigation and adaptation consequences.

Sources: Appendix 1 (40); Caldecott *et al.* (2014: 17, projects 7F-03786 in Lao PDR, 7F-04289 in Lao PDR, Tanzania, Madagascar, Indonesia and Cameroon, 7F-05222 in Lao PDR, Tanzania, Madagascar, Indonesia, and the transboundary zone of the Congo Republic, Cameroon and the Central African Republic, and 7F-06872 in Macedonia).

Community-based conservation

The community-based resource management (CBRM) approach is applicable in principle to any kind of renewable natural resource (RNR), including the full range of products and processes that are available from natural ecosystems, from timber and non-timber forest products (NTFPs) to medicines, foodstuffs, food additives, cosmetics, and ecotourism and educational opportunities, as well as water catchment and carbon storage services. The basic idea is that if improvements in the livelihoods of local people are based on their obtaining a share of the wealth generated by RNRs, then they will want to protect them and resist attempts by outsiders to destroy them. The whole approach is rooted in traditional practices in many locations, in which local people lived as RNR user groups in their home environments, managed their own RNRs according to traditional rules, and defended them against outsiders. These systems worked well in stable conditions in the past, but many have broken down under modern pressures, such as where nomadic peoples are settled and forest-dwelling peoples are deprived of their forest rights. Noting "the lack of donor recognition of traditional knowledge and systems", some argue that "one of our priorities should be to instigate those systems again as the central core for any environmental based work, and then build out from there", with Timor-Leste being cited as an example of a country that "is very supportive of re-establishing traditional law (*tara bandu*) and has legislated the requirement for a collaborative management model between government and communities" (Trudiann Dale, Conservation International Country Director, Timor-Leste, personal communication, May 2016).

This attitude can perhaps be traced to a re-thinking of the community role in conservation that was particularly inspired by the Communal Areas Management Programme for Indigenous Resources (CAMPFIRE) in Zimbabwe in the 1980s (Martin, 1986; Kiss, 1990; Child, 1996). Influential because of the high profile of wildlife and hunting tourism, this evolved into the communal conservancies system in Namibia in the 1990s (Jones, 2010), and similar arrangements elsewhere in Africa (Hulme & Murphree, 2001; Forslund, 2007; Jones, 2010; Nelson, 2010). It has also been applied to forest management in Cameroon (Gardner, 2002) and proposed for charcoal production in Village Land Forest Reserves in Tanzania (Caldecott *et al.*, 2013). A similar arrangement, originally conceived for wildlife but later applied to all community-owned resources, is the Community Resource Management Area (CREMA) system in Ghana (Living Earth, 2008; Asare *et al.*, 2013; IUCN, 2015), in which landscapes are defined by social boundaries within which (often for traditional reasons) there is social cohesion and a basis for collective action, matched where possible to ecological boundaries, legally defined, and authorised through devolved powers to manage all local natural resources in the interests of CREMA members, with revenues obtained and shared locally from their sustainable use. The observed affects have included increased conservation awareness, wildlife species returning and populations increasing, forest re-growth, reduced bushfires, and new livelihood opportunities.

This is not, however, just a matter of pointing out a product and a market for it, and facilitating links between them, but also requires ownership and distribution issues to be resolved. Thus, for this kind of approach to work, real powers must be transferred unequivocally to local management committees and communes, including powers to manage RNRs and benefits from them, and to make decisions about local management. These transfers may be resisted by other interest groups (Oyono, 2004), and this resistance may be heightened where attempts are made to apply the same principles of local benefit capture to non-renewable resources such as petroleum and hard-rock minerals. Nevertheless, experience across Africa and elsewhere does suggest that the CAMPFIRE-like approach works well provided that an acceptable distribution of ownership and revenue can be negotiated and implemented in a transparent way, and that revenue streams and their distribution are adequate to offset the marginal value to an individual of cheating the system (Caldecott, 2005). Even small amounts per person can be enough to change attitudes, if they are regular and come to be expected so that future earnings can be factored into decisions such as whether or not to set a snare in a particular place. Problems arise when non-local stakeholders take too large a share of the earnings, for example those from gorilla tourism being used to finance a whole park system, or when the source of revenue dries up, for example as a result of armed conflict. Even so, once a revenue stream from a wildlife resource has been experienced, attitudes to that resource may be fundamentally changed and local people may still consider it worth keeping in the hope that revenues from it will one day be restored.

Similar experiences are applicable to RNRs that form part of an area that has been alienated from local control (but not from vulnerability to local exploitation), for example a conservation area or national park, or a private ecotourism resort. These matters have been explored widely since the 1980s, often at a governmental level with the support of official donors through integrated conservation-development projects (ICDPs), in which certain benefits are offered to local people in exchange for their cooperation in conserving the target area, and these can include a share of protected area revenues (entry fees, overnight charges, etc.). The main challenge for either an ICDP-like or a CAMPFIRE-like livelihood-based conservation strategy lies in achieving governance that is good enough to support just and durable arrangements, including a long-term arrangement that solidifies a social contract in the minds of all concerned.

For ICDPs, the underlying transaction is that the costs imposed on local people by having a protected area in their midst are offset by the benefits offered in the form of livelihood support by or on behalf of a conservation agency. For CAMPFIRE-like systems, the costs imposed by tolerating wildlife are offset by earnings from wild species, mediated by some publicly accountable institution such as an NGO, trust fund, or elected community body. If they are to work properly, these exchanges must be clear, consistent, and trustworthy, and will be undermined by corruption, violence, or favouritism. In both kinds of enterprise, local people must have sufficient authority to negotiate freely enough to protect and advance their interests. The ultimate expectation is that the ecosystems concerned will be safer than would otherwise be the case because local people and businesses depend on their sustainable use, and/or because they generate revenues for the state. Whether the expectation bears fruit, however, depends on how carefully numerous factors have been considered in designing and planning the initiative, and mistakes do sometimes happen (Box 6.4).

Box 6.4 Lessons from ineffective biotrade projects

Biotrade-based conservation projects investigated in South Africa and Perú were scored 1 and 2 respectively for mitigation effectiveness. In South Africa, this low score was because the target tree species (the African baobab, *Adansonia digitata*) was considered 'sacred' and according to local informants would never be felled regardless of the intervention. In Perú, it was because although some of the species involved were forest-dwelling, others required non-forest habitat, and there was a lack of evidence for climate change effectiveness. A conclusion was that while biotrade-based conservation is a valid concept, projects based on it have very specific design requirements if they are to be effective.

Sources: Appendix 1 (40); Caldecott *et al.* (2014: 139–142, 231–233, projects UR-00568 in South Africa and UR-00276 in Perú).

Bioprospecting

At the high-technology end of the biotrade spectrum lie those concerning bioprospecting, the search for commercial products and processes using biodiversity as a source of information and as a stimulant for new ideas. The species richness of many tropical ecosystems offers a reservoir of biological information, and attendant biochemical richness, that is huge in scale but very poorly known. Chemicals found in nature have often yielded products with commercial value, including pharmaceuticals such as the quinine-based anti-malarials, cocaine-based anaesthetics, aspirin-based analgesics, and antibiotics including penicillin and erythromycin. But systematic research to identify commercial products from natural sources has only developed since the 1980s with the growth of biotechnological techniques and computing power (ten Kate & Laird, 1999). This began with a focus on tropical rainforests (e.g. Caldecott, 1987; Reid *et al.*, 1993), but has since extended into marine environments (e.g. Bhatia & Chugh, 2015; Table 6.1; Box 6.5).

Table 6.1 Marine natural products in the health and beauty markets

Category	*Product (application)*	*Source organism*	*Status*
Therapeutic	Yondelis® (cancer)	*Ecteinascidia turbinata* (ascidian)	approx. €60 million in 2012
Therapeutic	Prialt® (neuropathic pain)	*Conus magus* (mollusc)	est. US$20 million in 2012
Therapeutic	Halaven® (cancer)	*Halichondria okadai* (sponge)	est. >US$200 million in 2011
Therapeutic	Salinisporamide (cancer)	*Salinispora tropica* (bacterium)	Phase I
Therapeutic	Plinabulin (cancer)	*Aspergillus* sp. (fungus)	Phase I
Biofilm inhibitor	Brominated furanones (quorum sensing inhibitor)	*Delisea pulchra* (red alga)	In trials
Sunscreen	Mycosporine-like amino acids (UV absorbing)	Coral zooxanthellae	In trials
Cosmetic	Pseudopterosins (anti-inflammatory)	*Psuedopterogorgia elisabethae* (soft coral)	Commercial
Cosmetic	Venuceane (anti-free radicals)	*Thermus thermophilus* (bacterium)	Commercial
Nutrition	Omega-3 fatty acids	*Crypthecodinium cohnii* (microalga)	Commercial
Nutrition	Carotenoids (anti-oxidant)	*Dunaliella salina* (microalga)	Commercial
Therapeutic	Finafloxacin (fluoroquinolone antibiotic)	Microbial source	Approved by US FDA, Dec 2014

Source: Global Ocean Commission (2013); FDA (2014).

Box 6.5 Bioprospecting in marine ecosystems

"The marine realm contains a very rich variety of organisms, many of which remain undescribed. Because of their high biological diversity, marine ecosystems are particularly suited for bioprospecting, a process that aims to identify and isolate natural compounds from genetic material. Today, about 18,000 natural products have been reported from marine organisms belonging to about 4,800 named species. The number of natural products from marine species is growing at a rate of 4% per year. Since 1999, the number of patents of genetic material from marine species has increased at the rate of 12% per year. Marine species are about twice as likely to yield at least one gene in a patent than their terrestrial counterparts. Even this is likely to be an underestimate, because cloning and sequencing techniques allow description and patenting of genes of species yet to be named or even discovered. The applications of genes of marine organisms cover a wide range of activities, including pharmacology and human health, agriculture, food, cosmetics and industrial applications. However, it is in the area of pharmaceuticals that there has been most public interest. The success of natural compounds in drug discovery is unparalleled: for antimicrobial and anticancer therapies, for example, more than 70% of new chemical entities introduced during the period 1981–2002 originated from natural products. It has been estimated by the US National Cancer Institute (NCI) that 1% of samples from marine animals tested in the laboratory reveal anti-tumour potential (which compares favourably with just 0.01% of samples of terrestrial origin). This impressive rate of discovery is not likely to slow down. To date, sampling of marine products has primarily occurred in easy-to-reach coastal waters. As a result, 97% of natural products of marine origin are from eukaryotic sources (organisms with complex cells), with sponges alone accounting for 38% of the products. However, the majority of the Earth's metabolic diversity resides in prokaryotic organisms (single celled organisms such as bacteria) and over 99% of the microbial community of the ocean remains to be explored."

Source: Global Ocean Commission (2013: 1–2).

Whether in forests, soils or coral reefs, there is the possibility of using traditional knowledge held by indigenous people, or scientific knowledge held by biologists, to guide bioprospectors towards items that might be expected to have particular kinds of potency, or of randomly screening large numbers of naturally occurring biochemicals for particular kinds of activity. In all cases there are intellectual property and/or sovereign ownership issues that need to be addressed through free, prior and informed consent and mutually agreed contracts that cover all eventualities including up-front payments in cash or kind, and the sharing of any revenues that may result from successful commercialisation. These issues have been addressed in international law through the CBD and its protocols, and through national laws and contracting practices negotiated by governments, peoples and companies over several decades.

A regional biodiversity initiative

The Association of Southeast Asian Nations (ASEAN) has built some shared institutions to promote cooperation, including in the biodiversity sector (Box 6.6). In response to a regional biodiversity crisis (Box 6.7), the EU made a grant to ASEAN in 1999 with which to establish the ASEAN Regional Centre for Biodiversity Conservation (ARCBC). The story that then unfolded is replicable in part, since it sheds light on problems surrounding the loss of biodiversity and adaptation capacity at a regional level, their resistance to being solved, why countries may wish to cooperate in solving them, and how some progress at least can be made. The understanding that emerged of the nature of the biodiversity sector, the value-adding roles of regional institutions, and the centrality of biodiversity to climate change adaptation, are all of universal relevance.

Box 6.6 ASEAN and its biodiversity sector

"The main vehicle for regional co-operation in South-east Asia is the Association of Southeast Asian Nations (ASEAN), an inter-governmental organization that was founded in 1967 to promote 'peace, freedom and prosperity' (*Bangkok Declaration*, 8 August 1967). Its founding declaration included a commitment to inter-generational equity ('posterity'), which is a central concept in conservation, and ASEAN has developed a number of mechanisms to promote regional co-operation on environmental and biodiversity issues. By the mid-1990s, the concept of establishing an ASEAN institution to promote knowledge sharing about best practices and common efforts in the biodiversity sector had arisen, and led to a proposal for EU collaboration in establishing an ASEAN Regional Centre for Biodiversity Conservation (ARCBC).

"There is an emerging awareness of a biodiversity sector that has distinctive investment needs and attributes, while also being extremely large, complex and interactive with other sectors. This sector can be defined to include everything to do with:

- *saving biodiversity* (e.g. the design, protective management, financing, use, planning, staffing and inter-sectoral significance of protected areas and protected area systems; the management of genetic resources, species, populations and ecosystems outside protected areas; the

suppression of alien invasive species, fires and other factors that pose a threat to wild species populations; and relevant legislation and policies);
- *studying biodiversity* (e.g. all actions to do with research and inventory work involving the collection of information of any kind related to any aspect of genetic resources, species, populations and ecosystems and the organization and use of that information);
- *teaching about biodiversity* (e.g. all actions that use information about biodiversity for an educational purpose whether commercialized or not); and
- *using biodiversity* (e.g. in agriculture, medicine, bioprospecting, ecotourism, natural history film-making and journalism).

"These four dimensions of the sector respectively concern the survival of the raw material upon which it is based (and vital ecological services, such as stable water supplies and avoided floods and droughts), the expansion and organization of our knowledge about it, the stimulation of our minds to think creatively about how to use it, and the application of our energies to creating wealth from it."

Source: Caldecott *et al.* (2004: 4–5).

Box 6.7 The biodiversity crisis in ASEAN

"The South-east Asian region is of immense importance for global biodiversity, since it contains the natural habitats of up to 40% of all species on Earth. It includes three 'mega-diversity' countries (Indonesia, Malaysia and the Philippines), several biogeographical units (e.g. Malesia, Wallacea, Sundaland, Indo-Burma and the Central Indo-Pacific), and numerous centres of concentration of restricted-range bird, plant and insect species. Species richness by area is higher in several ecosystem types (e.g. lowland rain forest, coral reefs) than anywhere else on Earth, and overall species richness is known to be very high although most species are little-studied invertebrates and unknown to science. These biodiversity resources have evolved and exist in the context of natural ecosystems, and cannot survive if those ecosystems are destroyed" (Caldecott *et al.*, 2004: 4).

The scale of the biodiversity crisis in ASEAN is immense and continuing. A potent indicator of the processes that have been underway, and their consequences, is the loss and condition of natural forests. Natural forest cover has been catastrophically reduced in the Philippines, and widely lost, fragmented and/or degraded in all the other countries, with Indonesia and Myanmar in particular experiencing rapid rates of deforestation. Causes of forest degradation and loss include:

- *logging* (reported as a cause of deforestation in all the countries – logging typically starts legally, then becomes illegal when logging is banned after serious environmental problems emerge, as occurred in Cambodia, Lao PDR, the Philippines, Thailand and Vietnam; another factor is that logging may be driven by demand in other countries, as with demand in India, China and Thailand affecting the forests of Myanmar, and in Malaysia and China driving logging in Indonesia);
- *fuelwood collection* (in Myanmar and Cambodia);
- *shifting cultivation* (in Indonesia, Lao PDR, the Philippines, Thailand and Vietnam);
- *mangrove-replacement aquaculture* (in Myanmar, Indonesia, the Philippines and Thailand);
- *forest fires* (in Indonesia and Malaysia);
- *development of tree plantations* (in Myanmar, Indonesia, Lao PDR, Malaysia, Thailand and Vietnam); and
- *aggravating factors*, such as:
  - timber smuggling (e.g. from Myanmar to China, India and Thailand, from Indonesia to Malaysia and China, and from Lao PDR to Vietnam);
  - hunting and harvesting non-timber forest products and the smuggling of resulting products (e.g. from Myanmar, Lao PDR and Vietnam to China);
  - alien invasive species (especially in Myanmar, the Philippines and Thailand), which may include the exotic trees being used to reforest large areas of Lao PDR; and
  - weak forest governance and law (conflicting policies, overlapping jurisdictions, lack of awareness and communication among stakeholders, etc., all widespread).

Other ecosystems have also suffered extensive damage in these countries, including:

- *coral reefs*, extensively damaged by destructive fishing, coral mining, pollution and sedimentation (e.g. in Myanmar, Indonesia and the Philippines);
- *wetlands*, extensively damaged by drainage, dams, pollution, construction and farming (e.g. in Thailand), or by 'ghost drainage' and fire (e.g. in Indonesia); and
- *karst (limestone) areas*, destroyed by quarrying (e.g. in Indonesia, Lao PDR, Thailand and Vietnam).

Sources: Caldecott *et al.* (2004); Caldecott & Pagett (2006).

The performance of the ARCBC was evaluated in 2002, focusing on its key functions of networking, training, research and database management, for all of which it was given low to moderate scores in relevance, efficiency, effectiveness, impact and sustainability although the trend was generally towards higher performance (Table 6.2). At the time there was limited enthusiasm for the institution among some of its stakeholders, but a follow-up investment was nevertheless recommended, and a financing proposal for a successor institution was developed (Caldecott *et al.*, 2004). The ASEAN Centre for Biodiversity (ACB) was then established, monitored in 2007 and early 2008, and evaluated more thoroughly in late 2008, when it earned high performance scores across the board (Table 6.2).

At that point, in light of continuing reluctance among some stakeholders, a number of contradictory expectations of the ACB were highlighted, and effort was made to clarify what can and cannot be expected of an institution of this type (Box 6.8). A diversified funding strategy focusing on links between biodiversity and climate change was recommended (Box 6.9), and, when EU support ended in 2010, the German government stepped in as the main donor. Since then, it was reported in January 2016 (Berthold Seibert, personal communication), the ACB had enhanced its portfolio on biodiversity and climate change, the practical economics of ecosystems and biodiversity, agricultural value chains in the buffer zones of protected areas, institutional strengthening across the biodiversity sector in ASEAN, and support to the ASEAN Heritage Parks initiative, which now encompasses 37 protected areas in all ASEAN countries. A new EU grant was being prepared, and the ACB was obtaining work contracts from ASEAN countries and international organisations, while facilitating ASEAN participation in the CBD and networks such as the Intergovernmental Platform on Biodiversity and Ecosystem Services.

Table 6.2 Performance of the ARCBC and ACB, 2002–2008

Criterion	*Score by year*				
	2002	*2004*	*2007*	*May 2008*	*Nov 2008*
Relevance	3.5	4.5	5.0	–	6.0
Efficiency	3.5	4.0	3.0	3.0	5.0
Effectiveness	4.0	5.0	3.0	3.0	5.0
Impact	3.0	5.0	3.0	5.0	5.0
Sustainability	3.5	3.5	5.0	5.0	5.0
Mean score	3.5	4.4	3.8	4.0	5.2

Sources: Caldecott *et al.* (2002) for 2002 and 2004; Rodinson & Felicisimo (2007) for 2007; Embrechts (2008) for May 2008; Caldecott & McNally (2008) for November 2008.

Box 6.8 Roles of the ASEAN Centre for Biodiversity

"The ACB was explicitly designed as a small but strategic institution to help the ASEAN member countries cooperate on biodiversity issues that can only, or can best, be addressed at the level of the region (i.e. all of ASEAN, or ASEAN and its neighbours) or sub-region (i.e. more than one member country). These include such challenges as:

- managing trans-frontier reserves and migrating wildlife populations;
- addressing trans-frontier wildlife and timber trade issues;
- conserving river-system, river-basin, coastal and marine ecosystems;
- addressing root causes of forest fire and smoke pollution;
- promoting understanding of governance and legislative features that encourage and enable conservation;
- establishing common professional competency standards for conservation tasks such as managing protected areas and resisting alien and invasive species; and
- protecting members' common interests in biosafety, bioprospecting and resisting biopiracy.

"Thus the ACB was not designed as a regional conservation agency, which might be expected to deploy resources to resist conservation problems in particular locations. This is a different role, and one best performed by individual governments with the support of international NGOs and official donors. By contrast, the ACB was designed to use the experience of member countries in attempting specific conservation tasks, to identify common themes, lessons learned and best practices, and then to ensure that all the members have access to this information. Similarly, the ACB was not designed do research on biodiversity, which is the province of each member country's own research community and its international partners, but rather to add value to collective knowledge holdings. It was designed to do this by ensuring that biodiversity information is shared among the members, and by using the information resource to help them meet their own particular needs, whether this is to fulfil treaty obligations on biodiversity reporting, or developing policy on matters of special interest to each country, or training staff. From this it can be seen that ACB was designed as an institution to manage knowledge on behalf of the ASEAN member countries collectively, and to build capacity within each one where this can best be done by sharing skills and pooling knowledge.

"Such a role was always likely to generate contradictory expectations. The conservation needs of the ten ASEAN countries are considerable, and the demand for resources to address them in most cases exceed national budgets and aid flows. Besides which, there are always important issues that fall between, or are not yet being addressed by, existing budget lines and projects. The case for a regional conservation funding agency is therefore unanswerable, but ACB was never intended to be it. The GEF might reasonably have financed such an institution with global biodiversity funds, and another opportunity may now be approaching with financing for climate change adaptation and mitigation. But the ACB was designed with the much more limited intention of filling a key strategic gap in the ability of member countries to learn from each other (and to an extent from the international community) and to build common positions around solutions to biodiversity challenges and opportunities.

"The net result is tension between countries that seek practical, problem-solving field activities in their own territories, and an institution with a different and more strategic purpose. This is analogous to a situation in which there is public demand for a curative medical system, while what has been established is a preventative health service. All may agree that 'prevention is better than cure', but the demand for curative intervention will remain. The problem is that both are needed, but in the case of biodiversity, ASEAN still only possesses one of them."

Sources: Appendix 1 (6); Caldecott & McNally (2008: 15–16)

Box 6.9 Opportunities for the ASEAN Centre for Biodiversity

"The ACB got off to a slow start, partly due to an over-dependency on the Governing Board for key decisions, prior to the entry into force of the ACB Establishment Agreement. Since the ACB is already written into an annex of the ASEAN Charter, there is reason to expect it to be a persistent institution if not necessarily a permanent one. How well resourced it will be is another question. Biodiversity as a topic has declined in attractiveness to governments in recent years, despite mounting evidence of global mass extinction. Problems of biodiversity loss are complex and intractable, and in any case are being eclipsed by the need to respond to and mitigate climate change, with all its associated issues of human environmental and livelihood security. There will be an increase in biodiversity discussions in 2010, to mark the passing of the global target for reducing the rate of biodiversity loss in that year, but this is likely to be brief. It does however represent an opportunity for ACB to make the case for additional long-term funding, on the grounds that:

- problems of biodiversity loss have not yet been solved, so the ACB's services are essential and will be needed for the foreseeable future;
- the failure of the international community to reach the 2010 target

contains lessons that should be learned and the results disseminated over coming years; and

- even though insufficient, the 2010 target did draw attention to biodiversity loss and ASEAN will need to participate in trying to reach any new target that may be set.

"Meanwhile, though, the next few years will see a rapid increase in funding for climate change adaptation (including disaster risk reduction) and mitigation (including REDD). The future of the ACB, therefore, is partially tied in with its ability to find for itself a useful niche within the climate change debate. This is entirely feasible, since biodiversity is an attribute of ecosystems, and viable ecosystems are essential to human interests. Knowledge about managing biodiversity is largely transposable to the management of ecosystems as carbon sinks and stores, and as water catchments, with the added advantage that diverse ecosystems offer a greater range of services and are therefore more attractive investment targets than simplified ones. Thus the issue for the ACB is to articulate what it does, as a regional inter-governmental knowledge management and capacity building institution, in terms that those interested in human security can understand. This argument can be made purely in terms of biodiversity (such as obligations under the CBD and other environmental treaties), but can also be amplified in the context of climate change."

Sources: Appendix 1 (6); Caldecott & McNally (2008: 31).

It seems that the ACB has survived the challenges involved in its establishment, and the recent information suggests that the institution has a set of roles that are increasingly seen as useful by the ASEAN member countries and globally. Whether this can yield transformative outcomes remains to be seen, given the scale and momentum of environmental change in South-east Asia – which must include, for example, the real prospect of losing all remaining rainforests to drought and fire, and all remaining coral reefs to ocean acidification. But it is clear that the ACB is also highly replicable in the sense that all groups of nations facing common and inter-connected environmental problems have a need individually and collectively to meet their treaty obligations, and an ACB-like institution can assist them greatly in doing so. Other lessons learned from the whole process since the mid-1990s are that it is worth building on any cooperative tendency existing in a group of countries, but that it is essential to be realistic about who will pay for what and explaining why they should, and that a correct analysis of need and opportunity deserves long-term commitment in response, where necessary maintained by a degree of faithful optimism.

Payments for ecosystem services

Arrangements for ecosystem services to be paid for routinely as a way to enable investment in their management and continued availability is a key part of the solution to global problems. This is potentially applicable to all ecosystem goods and services, including carbon storage (the 'payment by results' dimension of REDD+ is essentially a PES arrangement), but particularly to water, either at a landscape level (Box 6.10) or more strategically (Box 6.11; also Box 6.20). Although such arrangements involve money (e.g. the payment of bills by water users), the people who manage the ecosystems that provide the services do not necessarily wish to be paid in cash. A PES arrangement should include all parties agreeing how rewards will be delivered to service providers, from among such options as grants or aid-in-kind for community-defined social activities or development projects, educational scholarships, jobs as catchment wardens or ecosystem restorers, or by sharing out cash.

Box 6.10 Payments for ecosystem services (PES) in China

Studies commissioned by a Conservation International project with its local partner ShanShui ('Mountain-Water') mainly focused on opportunities for PES interventions, and on conservation agreements for behaviour change and tenure reform on collectively owned forest land. The PES work led to arrangements whereby water fees (RMB¥300,000 as seed money and ca RMB¥750,000 recurrent, in cash) were being collected in Pingwu county town and paid via the Sichuan provincial government into a fund for managing catchment forests through community conservation. Other potential income streams were also being investigated by the project, including deals with a local hydroelectricity company, and carbon storage potential. An official openness to making PES arrangements based on user fees and covering catchments that also contain high levels of biodiversity (and flagship species such as giant pandas and golden monkeys) offered a potent model for replication across China.

Sources: Appendix 1 (9); Caldecott *et al.* (2009).

Box 6.11 PES in Costa Rica

Overview. The PES programme in Costa Rica is a national payment programme for carbon storage, hydrological services, and the protection of biodiversity and landscapes. It has been credited with reducing one of the world's highest rates of deforestation, and reversing it to net reforestation by the start of the 2000s. Between 1997 and 2004, approximately US$200 million was invested in PES to protect over 460,000 hectares of forests, to establish forestry plantations and to provide additional income to more than 8,000 forest owners. The PES system is managed by FONAFIFO, a semi-autonomous agency.

Financing. The PES system has mostly been financed by a 3.5 per cent share of revenues from a sales tax on fossil fuels, but the aim is that all beneficiaries of environmental services eventually pay for those they receive. While there has been some success in charging water users for upstream catchment management services, there has been more limited success in charging for biodiversity and carbon. The proportion of the programme's costs financed with direct payments will increase as a new water tariff is implemented.

Water service payments. It is intended that payments from hydroelectric power producers and other water users would at least partly finance PES. But as there is no legal requirement, FONAFIFO negotiates with water users and has reached a number of agreements. Progress was slow at first, but the process has been streamlined using environmental services certificates (standardised instruments that pay for conservation of a hectare of forest in a particular area). Both the number of agreements and the amounts paid have risen sharply. In the past, water users paid only one quarter of conservation costs (based on the idea that catchment management is one of four ecosystem services provided by forests), but recent agreements pay the full cost of conservation in addition to FONAFIFO's administrative costs.

Biodiversity payments. Biodiversity payments have mainly been financed by the Global Environment Facility (GEF). In contrast to agreements with water users, financing for biodiversity has been on a one-off basis, and efforts to make it sustainable by generating financing from tourism have not been successful. This is acceptable in areas where financing based on carbon or water management is possible, but there is a large area (roughly 900,000 hectares) which is a priority for biodiversity conservation, but which lies outside protected areas and does not have the potential for either water or carbon financing. An endowment fund is being created as a partial solution.

Carbon payments. While use of the fuel tax revenue can be considered to be a payment from carbon users to carbon conservers, because the tax is mandatory and because the revenue is used for payments to a range of environmental services, the link is rather weak. The aim has always been to sell carbon emission reduction credits, and PES contracts clearly state that FONAFIFO owns the right to emissions reductions. FONAFIFO developed the Certifiable Tradable Offset (CTO) equal to one tCO_2e of externally certified net reduction in carbon emissions. The programme sold 200,000 CTOs for US$2 million to the Norwegian government and a consortium of Norwegian power producers. No additional sales of CTOs have been made, however, as emissions reductions are mainly based on avoided deforestation and only reforestation and afforestation are considered eligible under the Kyoto Protocol's Clean Development Mechanism (CDM). Costa Rica has since sold 0.61 million tCO_2e to the World Bank BioCarbon Fund, based on a mix of planting trees in agroforestry systems, natural regeneration and commercial plantations. In order to deliver Kyoto-eligible carbon emission reductions more effectively, the PES system is introducing a new type of contract based on assisted natural regeneration.

Landscape payments. The Forest Law which provides the legal basis for the PES scheme mentions scenic beauty as an environmental service provided by forests. While there have been negotiations with hotels and a rafting company for payments for scenic beauty, no agreements have been reached. Users of landscape services are numerous and fragmented, and problems of collective action make implementing PES for landscape beauty difficult.

Impact on biodiversity. Despite the difficulty in charging for biodiversity services, because biodiversity is 'bundled' with other ecosystem services, there are significant biodiversity benefits associated with the conservation of catchment forests. The PES scheme has helped slow deforestation, added monetary value to forests and biodiversity, and increased understanding of the economic and social contribution of natural ecosystems. Identifying the specific results of PES schemes can be tricky when a number of policy changes are introduced simultaneously, and it has proven difficult to determine precisely the extent to which the PES programme has generated ecosystem services. While studies have found that PES recipients had a higher proportion of forest on their land than non-recipients, other research suggested that participants would have protected their forest even in the absence of the PES programme.

Replicability. The number of PES schemes in Latin America is growing, suggesting scope for replicability when there are clearly identifiable suppliers and intermediary institutions to facilitate payment. Costa Rica's PES scheme has been much studied and imitated. Many countries already have similar schemes in place, and these have often been used following policy reform including a shift from subsidies to PES.

Lessons learned. PES schemes are easier to introduce if they build upon existing systems of payments, and Costa Rica's system was based on reform of an existing forest subsidy programme. In the 1970s, Costa Rica had begun to provide incentives for timber plantations through tax rebates, due to concerns over shrinking timber supplies. The Forest Credit Certificate expanded the programme, which continued to evolve to support forest conservation as well as timber production. When introduced, the PES programme built on this payment scheme, with two major changes: (a) payments were to be based on the provision of environmental services instead of timber; and (b) financing would change from the government budget to an earmarked tax and payments from beneficiaries. While PES schemes are proliferating in Central and South America, it is proving more difficult to implement user-financed PES schemes than government-financed schemes, which limits their sustainability. Other lessons learned are as follows.

1 It is easier to implement PES for water catchments than for biodiversity (which is hard to measure) and for carbon (where it is hard to identify beneficiaries).
2 Effective targeting and differentiated payments are important to allow for differences in the level and opportunity cost of service provision.
3 PES schemes need to monitor and document how activities are generating environmental services, particularly for carbon conservation projects that aim to sell carbon offsets.
4 PES schemes should be flexible and able to adapt to lessons learned and changing circumstances – Costa Rica's experience has been broadly positive, yet its PES system continues to evolve in response to experience and feedback.

Sources: CBD (undated); TEEB (2010).

Grasslands and desertification

Pastoralism is central to the economies of many peoples living in major grassland areas, including the steppes of central and eastern Asia. Such a way of life depends on large-scale movements of people and herds, which spreads grazing pressure over vast areas. Where national boundaries, or cultural and economic changes, restrict these movements or encourage an increased stocking density of livestock, over-grazing is a real threat, and with it a loss of resilience and the risk of desertification. This danger can be managed by controlling grazing intensity through regulation of the animal stocking rate, by enhancing rotational grazing, and by limiting grazing time by season over the year (Box 6.12). Such strategies can lead to the storage of soil carbon to the extent of 0.11–1.50 tCO_2/ha per year (ADB, 2009), which makes the outcome relevant to mitigation and, potentially, to carbon markets (Box 6.13).

Box 6.12 Pasture ecosystem management and combating desertification in Mongolia

Pastoralism is central to Mongolian society, culture and economy, and 40 per cent of Mongolians earn a living as herders. Livestock-based range management continues to be their main productive activity and the land use with the greatest impact on environmental services in the country. Desertification already affects over 70 per cent of Mongolia's grasslands, and many stakeholders in Mongolia have come to believe incorrectly that climate change is the main cause of land degradation, rather than over-grazing due to unregulated and open access to pastures. The following three projects covered all three aspects of the dryland-overgrazing-desertification nexus.

- One project focused on equipping herders with a means to work with local authorities in collectively managing livestock in a sustainable way, which it did over some 21.7 million hectares of pastureland while also building increased community resilience to the consequences of climate change.
- Another project sought to coordinate national and international efforts for coping with desertification and promoting sustainable livelihoods in arid and semi-arid areas, by organising local communities around improved land management and the use of energy-efficient stoves, public awareness-raising and education, and open-access information on desertification coping techniques and best practices.
- The third project aimed to combat land degradation and desertification by strengthening coordination and capacity for sustainable land management, mainstreaming it into national strategies, policies and laws, and piloting the community-based management of grasslands and water.

The first and third parts of the portfolio were given effectiveness scores of 5–6, but the second appeared weaker probably because of the indirect nature of coordination and knowledge management. The three were nevertheless synergistic and likely to have long-term impact so collectively they were scored 5 for adaptation effectiveness.

Sources: Appendix 1 (40); Caldecott *et al.* (2014: 18, 150–154, 145–149, 165–168, projects 7F-03461, 7F-05405, and 7F-06465).

Box 6.13 Linking herders to carbon markets in Mongolia

The overall objective of this project was to reverse grassland degradation, improve rural incomes, and reduce herders' vulnerability to climate risk, by supporting adoption of sustainable grassland and livestock management practices and improved product marketing by herders. It aimed to do this by developing a pilot carbon finance project in which atmospheric carbon would be conserved in grassland soils through the adoption of sustainable grazing management practices, measured using methods that meet international carbon market standards, and ensuring that herders receive payments for the carbon sequestered. The project made no contribution to reducing GHG emissions, or the flow of climate finance to Mongolian partners, but it did contribute to important methodological work, and supported research and awareness raising that were relevant to both mitigation and adaptation. For a country like Mongolia, grassland management is critical to adaptation and combating desertification, and carbon sequestration has to form a central part of mitigation efforts alongside those focused on the mining and domestic heating sectors (as foreseen in the National Action Programme for Climate Change). While strategically relevant, the lack of specific progress led to the project being given an effectiveness score of 3.

Sources: Appendix 1 (45); Caldecott *et al.* (2014: 48, 169–172, project 7F-07809).

Farming systems

Food security

Investments in farming systems can include measures with aims such as the following:

- to manage knowledge and improve regulation and networks for the development and distribution of seed varieties with which to grow improved crops (i.e. those that are more nutritious and more productive under more diverse and challenging conditions; Box 6.14);
- to improve planting material for tree crops oriented to drought- and salt-tolerance; and
- to encourage and enable local stakeholders to understand climate change and find and share through networking, and/or to learn through training, new ways to improve the resilience and productivity of their farming and food storage systems and the soils and water-bearing ecosystems upon which they depend.

The common theme here is to remove knowledge-based, regulatory, or organisational barriers to the flow of potential solutions to livelihood constraints in the context of deteriorating environmental conditions (e.g. saline intrusion, drought, flood, and soil depletion) that are associated with or could be aggravated by climate change.

Box 6.14 Hill maize research in Nepal

The project aimed to increase maize supply in rural areas of Nepal through research on maize varieties, dissemination of chosen varieties to hill farmers, participatory variety selection, production of improved maize seeds, and linking farmers' feedback to policy decisions through farmers' assessments of the new varieties. The project also contributed to establishing a national research system and strengthening the capacity of farmer groups to produce maize seeds and deal with markets to obtain higher and more assured prices, and to building understanding of the importance of agricultural research at the national level. It also increased research capacity among national agricultural institutions. The maize varieties promoted through the project were resistant to drought, heat and lack of nitrogen and thus improved the adaptation capacity of communities, and the project was given a moderate adaptation effectiveness score of 4.

Sources: Appendix 1 (42); Caldecott *et al.* (2014: 23, 240–243, project 7F-03093).

Bamboo plantations

The woody grasses known as bamboos are among the fastest-growing tropical and sub-tropical plants, and the uses to which their strong woody tissues have been put by people are extremely diverse (Box 6.15). Prominent among them is their use as a construction material.

Box 6.15 Promoting bamboo planting, harvesting, processing, use, and sale

Two projects sought to promote bamboo as a pro-poor strategy, and in both cases there may have been some climate relevance as planting bamboo on degraded land could result in a net increase in woody biomass while stabilising slopes and regenerating soils against rain and storm damage. Substituting bamboo for cement or steel might also reduce the carbon footprint of buildings. The effectiveness would depend on the scale, the details of planting systems and sites, and the use of materials, and the projects were given a moderate mitigation effectiveness score of 4.

Sources: Appendix 1 (40, 46); Caldecott *et al.* (2014: 17, 299–302, projects 7F-04301 in Cuba and 7F-05697 in Vietnam and Lao PDR).

Disaster risk reduction

Overview

Disaster risk reduction (DRR) investments include those that promote the sharing of relevant knowledge, that establish centres of competence, and that analyse risk or plan for its management. The purpose is to integrate DRR within development planning at all sectors and levels of society, and to do so alongside capacity building through training and knowledge exchange. Other elements of a DRR mainstreaming portfolio might include the strengthening of decentralised disaster rescue systems, and the delivery of technical assistance and capacity building on the financial mitigation of sovereign disaster risk (i.e. catastrophe risk assessment, fiscal risk management of natural disasters, and capacity building for disaster risk financing strategies).

Protection and early warning against specific threats

Projects in a DRR portfolio (Boxes 6.16 and 6.17) might include arrangements: to build public awareness of potential disasters and what to do about them (e.g. through interest groups and public discussions, and agreeing roles and responsibilities in advance); to prepare for the impact of particular hazards (e.g. stockpiling resources, preparing escape routes, building shelters, and converting public buildings into disaster refuges); or to strengthen localities against known hazards (e.g. by restoring beach dune systems to absorb storm energies, or forest cover on deforested hillsides to reduce the risk of mud-slides). Another important theme is to set in place systems for providing early warnings of:

- *droughts*, which pose a complex challenge for early warning because of their slow onset and the complex interplay involved among water, soils, seasons, crops, aquifers, farm prices, land ownership, and other factors;
- *storms*, including storm surges, which are the most dangerous features of powerful cyclonic storms when they make landfall, being up to several metres deep, moving fast and turbulently, and often topped by vast waves;
- *floods*, which are a challenge to anticipate because flood risks have complex drivers that include the integrity of catchment ecosystems, land use in floodplains and the extent of canalisation of rivers, as well as the location, intensity and duration of rainfall;
- *fires*, which become a major threat when normally fire-resistant ecosystems are drained, or damaged, and/or subjected to unusual droughts, especially where people set fires to clear land for farming;
- *heat-waves*, which require coordinated action by local government, especially on water and electricity supplies, park opening hours, and working and teaching schedules, to protect whole communities and particularly the young, old, and sick (Shah *et al.*, 2014);

- *cold-weather disasters*, which require survival and livelihood impacts to be buffered and national policy shortcomings to be corrected (Box 6.16); and
- *remote geohazards*, including glacier lake outburst floods and landslides, in which a response is often needed to the consequences of melting glaciers and tectonic instability in mountainous terrain, so must involve a mixture of geological and hydrological investigation, good engineering practices, capacity building, and land use regulations.

Box 6.16 Disaster relief and prevention in Mongolia

The project responded quickly and effectively to meet the most urgent livelihood needs of herders affected by extreme cold and livestock mortality during the 2010 cold-weather disaster (*dzud*). The project also aimed to mainstream *dzud* responses based on practical experience and contribute to improved *dzud* preparedness and policy in the future, including by addressing factors that drive over-stocking and over-grazing. Although effective in disaster relief, and appreciated by the Mongolian partners, its adaptation effectiveness remained low, as no explicit measures were taken to analyse climate variability or to forecast climate change impacts into the project's prevention and preparedness measures, so it was scored only 3.

Note: A *dzud* disaster occurred in Mongolia in the winter of 2015–2016, with average temperatures of –25°C (day) to –38°C (night) across the country (Reliefweb, 2016).

Sources: Appendix 1 (41); Caldecott *et al.* (2014: 21, 161–164, project 7F-07572).

Box 6.17 Managing natural disasters in Tajikistan

Muminabad district, with about 72,000 inhabitants, lies in the south of Tajikistan close to the Afghan border. The project sought to strengthen integrated disaster risk management and coping capacity of local government, civil society organisations and the population at large. Although climate change was not mentioned in the project design, the project contributed indirectly to strengthened preparedness and adaptation capacity. Strong local ownership of disaster prevention activities, and strong interest in energy efficiency and renewable energy solutions (including solar cookers, energy efficient stoves, and household energy saving with heat exchangers), seemed able to help reduce pressure on forests and soils in a hilly area 700–3,000 m above sea level, which are consistent with adaptation and the project was scored 6.

Sources: Appendix 1 (41); Caldecott *et al.* (2014: 21, 317–320, project 7F-02864).

Disaster risk insurance

These investments include measures to facilitate the compensation of disaster-related damage through insurance mechanisms, thus increasing socioeconomic resilience to the effects of climate change. The scale of the approach ranges from inter-governmental risk sharing in relation to macro-scale events such as region-wide droughts and floods, to promoting micro-insurance against disasters and other hazards for small-scale farmers and microcredit borrowers. Index systems may also be developed that exploit known correlations between environmental conditions and livestock mortality or crop losses, thus simplifying and speeding insurance claims (Box 6.18). Such measures require research to establish and quantify correlations (and/or to establish causality), and environmental monitoring to detect changes that would trigger claims. Insurance pay-outs are likely to support local adaptation efforts because the claimant has the opportunity both to learn from what went wrong (i.e. to understand and quantify vulnerability) and to 'build back better' (i.e. more resiliently, using capital to invest in more robust farming systems or housing, or to relocate to a safer place). This would apply at the micro- and macro-levels, including the national level where strategic adaptation decisions on major infrastructure and development zoning can be taken, and where aid (with donors acting, in effect, as underwriters) can be more efficiently deployed in response to calamity both for investment and humanitarian relief purposes. Moreover, the risk-sharing nature of insurance promotes awareness of hazards, incentivises investment in hazard reduction, and encourages social solidarity, which are all likely to be important in the face of climate change.

Box 6.18 Index-based agricultural insurance

Livestock insurance in Mongolia. The main aim of the project was to ascertain the viability of index-based livestock insurance, in order to reduce the impact of livestock mortality for herders from *dzud* and other events. The project was effective in reducing the impact of livestock mortality on herders, as well as reducing overall vulnerability to climate extremes. When reviewing the insurance premiums, the project also investigated forecasted implications of climate change in Mongolia. Although no major changes to the premium were considered necessary, this is a sign that the scheme was being explicitly climate proofed, and was actively addressing weather extremes as well as the challenges of advancing climate change, so it was given an adaptation effectiveness score of 5.

Crop insurance in southern Africa. The project focused on exploring the potential for developing index-based crop-yield insurance products and credit facilities in Zimbabwe, Swaziland, Zambia and Malawi, based on weather monitoring and pay-outs based on known correlations between water supply and the yields of various crops. Although delayed and imperfectly documented, the project was judged to be fully relevant to adaptation and able to help reduce the vulnerability of farmers and their families to weather extremes, so it was given an adaptation effectiveness score of 5.

Sources: Appendix 1 (41); Caldecott *et al.* (2014: 21–22, 155–160, 354–356, projects 7F-06642 in Mongolia and 7F-07807 in southern Africa).

Water resources management

Overview

The regular supply of adequate fresh water is a landscape-level challenge involving the functionality of catchment ecosystems (whether farmed or not) in capturing, holding and releasing water while retaining soils, interlinked downstream with the condition of streams, rivers, aquifers, and floodplains, and the distribution of houses and other infrastructure.

Area-based water resources management

Awareness and management of factors that can cause flooding, pollution and, in the case of irrigated farmland, salinisation, are important parts of an area-based water resources management (WRM) strategy. Among its hallmarks is a systemic approach to multiple interactive issues, built on a clear understanding of the area's ecology and around the comprehension and participation of the area's inhabitants (Box 6.19). Since understanding by stakeholders is so central to WRM, investments often focus on measures to enhance the management of information about water resources and how to manage them effectively and sustainably, including the provision of technical support in maintaining water systems and improving irrigation and drainage techniques.

Box 6.19 Area-based water resources management

Features of high-scoring adaptation projects. In Pakistan, promoting participatory village-based improvement in water retention and irrigation, water conservation, and catchment restoration; in Nepal, rehabilitating water-management infrastructure and raising awareness on the need for catchment management; in the Ferghana Valley of Uzbekistan, Kyrgyzstan and Tajikistan, promoting ecological awareness among water users and establishing cross-border cooperation in irrigation and disaster preparedness; and in Morocco, promoting re-use, multiple use, and more efficient use of water resources, integrated water use plans, and sustainable rain water management.

Features of low-scoring adaptation projects. In Tunisia, Egypt and Mali, a relative lack of attention to the social and ecological dimensions of WRM and a more exclusive focus on water access and delivery, water use efficiency, and urban flood risk management.

Sources: Appendix 1 (43); Caldecott *et al.* (2014: 23–24, projects 7F-07815 in Pakistan, 7F-07757 in Nepal, 7F-00934 in the Ferghana Valley, 7F-08025 in Morocco, 7F-08076 in Tunisia, 7F-08368 in Egypt, and 7F-07764 in Mali).

Payment for water catchment services

A WRM portfolio may also contain investments to do with spreading knowledge about PES arrangements for water catchments, by collecting global lessons learned and best practices, and packaging and disseminating the information to various audiences (Box 6.20).

Box 6.20 Payment for water catchment services

A global project was designed specifically to promote payment for 'watershed' (i.e. catchment ecosystem) services. Its credit proposal noted that such transactions exceeded US$9.3 billion in 2008, and described the goal as being to contribute to the development and implementation of a new generation of larger scale PES arrangements that engage and benefit rural communities, that value natural infrastructure approaches to better sustain ecosystem health, and that act as a cost-effective front line defence against climate change. The aim was to collect global lessons and best practices, and package and disseminate the information to relevant audiences. The leverage potential of the project was recognised with a very strong adaptation effectiveness score of 6.

Sources: Appendix 1 (43); Caldecott *et al.* (2014: 24, project 7F-07992).

Physical and institutional rehabilitation of water systems

Investments in water system rehabilitation include measures concerned with the sustainability of those systems, such as:

- *setting standards and forming alliances* with government institutions, and building water-management capacity at local level;
- *establishing control systems* for large irrigation schemes;
- *promoting decentralised management* of drinking water;

- *supporting government* in developing the policy basis, design, and operation of water and sanitation systems; and
- *helping water companies* to place their businesses on a more stable and sustainable footing, by way of rehabilitated infrastructure and improved metering, billing, payment collection, and financial and operational management.

Such investments may involve changing the ways in which water is moved around over long distances, by rehabilitating and redesigning gravity systems and improving pump systems, which would have consequences for energy efficiency. They may also include the rehabilitation of small-scale irrigation systems, including ancient tunnel systems (known as *kahriz* in Azerbaijan, and *qanat* in Iran) which are an inherently sustainable way to harvest ground water (Box 6.21).

Box 6.21 *Kahriz* rehabilitation in Azerbaijan

Much of Azerbaijan lies in one of the driest regions on Earth, and the country has far fewer water resources per person than other countries in the South Caucasus. The project focused on enhancing communities' access to water through rehabilitation of *kahriz* systems in the rural areas of the Nakhchivan Autonomous Republic (an exclave of Azerbaijan separated from the rest of the country by Armenia). The rehabilitation of 42 such systems provided drinking and irrigation water to nearly 3,000 families, and allowed more than 200 ha of additional lands to be irrigated. The strengthening of local livelihoods (also implied by evidence of reduced emigration from the autonomous republic) and increased DRR preparedness (including an enhanced role for women in the Water Users Committees), were assessed as strengthening the adaptive capacity of the communities concerned, but the project's lack of explicit screening for climate change impacts reduced its adaptation effectiveness score slightly to only 5.

Sources: Appendix 1 (43); Caldecott *et al.* (2014: 24–25, 313–316, project 7F-02242).

Water diplomacy

The potential for friction over shared rivers is immense, given the large number of rivers and river basins that cross national boundaries. Events inside each country's share of the catchment change the river downstream, and can include deforestation and erosion, irrigation and agrochemical use, the discharge of toxic effluents, the escape of leachates from garbage dumps, and the release of untreated sewage, as well as the building of dams and canals, and the diversion of water to cities. The scope for dispute is so great that water diplomacy is an important job of foreign ministries in many countries. This may be recognised through investments that seek to promote collaborative regional solutions (Box 6.22), but many of the same issues occur within countries, as administrative boundaries often cross catchments and rivers and there is the potential for water use, dam-building, pollution, etc., in one area to impact people and businesses in another (Box 6.23). These conflicts must often be resolved at high level within the government concerned, using its planning, arbitration, and conflict resolution powers, for which various kinds of capacity building are essential.

Box 6.22 Hydrodiplomacy in the Middle East

The project aimed to establish a regional reference point for sharing knowledge about sustainable WRM as a contribution to peace-building, and supported decentralised water efficiency projects in the Orontes river basin shared between Syria and Lebanon. The approach was assessed as very challenging when it was conceived in 2010 (and the circumstances in Syria have deteriorated since), but its potential for good adaptation effectiveness meant that it was scored 5 on principle.

Sources: Appendix 1 (43); Caldecott *et al.* (2014: 25, project 7F-07689).

Box 6.23 A national water catchment plan in Bolivia

To address complex river basin issues within one country, the project aimed to build capacity in the Vice-Ministry and Interagency Water Council to implement the *Plan Nacional de Cuencas* (National Catchment Plan) through training and technical advice to support the development of policy and legislation, establishment of standards and learning networks, and in planning, execution, tracking, and monitoring of implementation. Promoting informed dialogue among stakeholders is essential to adaptation, so the project was scored 5 for potential effectiveness.

Sources: Appendix 1 (43); Caldecott *et al.* (2014: 25, project 7F-05631).

Environmental monitoring

Environmental monitoring investments concern the collection, analysis and dissemination or exchange of scientific information, whether focused on renewable natural resources research systems, changes in arid environments, aquifer systems, water catchments, glacier dynamics and glacial outburst floods (Box 6.24), or biodiversity. The common theme is the use of knowledge to improve the quality of more applied actions, such as planning to reduce risks and improve early warning. The utility of public science focused on environmental change lies in its ability to increase public understanding of climate change, which can fuel voting preferences and activism in favour of mitigation and/or adaptation efforts. Thus information flowing from environmental research and monitoring can help focus and target adaptation and mitigation efforts, and allow decision makers to demonstrate that they are responding to public concerns.

Box 6.24 Early warning of glacier lake outburst floods in China

Melting ice from receding glaciers in Western China is heightening the risk of glacier lake outburst floods (GLOFs) to critical levels. The Yarkant River is located in the southwest of Xinjiang Province, at the margin of the south-western Tarim Basin, and ranks first in Xinjiang in flood frequency and in losses caused by floods that threaten an alluvial area of about 50,000 sq. km. Implemented under the umbrella of broader WRM actions (e.g. on dam safety and flood management), the project supported: flood modelling and climate monitoring and analysis (including compilation of base maps and establishment of a detailed digital elevation model of the Kyagar Glacier Lake basin, and evaluation of future scenarios for Kyagar GLOFs considering global climate change); early warning system development and establishment (including satellite remote sensing for early warning of GLOFs, installation of gauge and warning stations, web cameras, etc.); and increased understanding of glacier change processes under conditions of climate change. The GLOF early warning system was made fully automatic, based on defined thresholds for triggering alarms, and there was evidence of capacity building, training, and technology transfer, matched by a commitment of Chinese partners to make efficient use of deliverables. The project also generated knowledge about glacier development that is relevant to similar projects elsewhere (including those supported by the same donor in Perú). It was therefore scored 6 for adaptation effectiveness.

Sources: Appendix 1 (44); Caldecott *et al.* (2014: 26, 346–349, project 7F-07733).

Policy development

Policy investments apply the principle that solutions to climate change issues can best be found, improved and applied (whether through laws, plans, strategies, or practical actions) through informed dialogue among concerned stakeholders (Boxes 6.25 and 6.26). Such investments may promote knowledge sharing on climate risks and solutions, or mainstreaming adaptation measures into development decisions, or raising climate awareness among decision makers. The entry point can vary among WRM, forest ecosystem management, or the strengthening of national or local government planning in relation to adaptation and/or mitigation. They may also be designed to promote:

- *the flow of knowledge* about environmentally sustainable development, between cities and among rural areas or institutions;
- *legislative collaboration* between countries on clean air and low-carbon development options;
- *the use of economic criteria* to inform adaptation decision making; and
- *sustainability standards* around the environmental impacts of commodity markets and global product chains.

Box 6.25 An adaptation programme in Perú

The *Programa de Adaptación al Cambio Climático* (PACC) supported the development of a diagnostic tool for assessing climate change vulnerability in water catchments in Cusco and Apurímac. It compiled local knowledge relevant to adaptation practices with the participation of local authorities and civil society, and piloted field activities based on it. This helped build improved understanding of adaptation processes and informed the national adaptation strategy and negotiation positions from the UNFCCC CoP 18 (2012) onwards. The project showed that it is of fundamental importance to include climate change scenarios, vulnerabilities, risk management strategies, local actions for adaptation, etc., as well as short-, medium-, and long-term indicators, right from the beginning of project design, since adaptation is a process that extends far beyond the duration of 'normal' projects. The project was given an adaptation effectiveness score of 6.

Sources: Appendix 1 (44); Caldecott *et al.* (2014: 27–28, 221–227, project 7F-05409).

Box 6.26 Mainstreaming adaptation knowledge in China

The project aimed to feed climate science into adaptation planning at the national and provincial level, and in multiple sectors. It helped to establish a multi-disciplinary research team in China capable of delivering data on climate change impacts to serve as basis for vulnerability and climate risk assessments. It also promoted regional (e.g. in Mongolia) and global knowledge sharing in support of policy-oriented climate risk assessments and adaptation planning, and in engaging stakeholders and informing national adaptation policy processes. The project was scored 6 for adaptation effectiveness, but was also considered likely to help China's policymakers understand the importance of mitigation in safeguarding national economic and social development.

Sources: Appendix 1 (44); Caldecott *et al.* (2014: 27–28, 342–349, project 7F-06983).

The idea that underlies policy development is that good policies are important, and they can be found by enriching and exchanging relevant knowledge and creating influential forums where they can be discussed and decisions made. An additional step is the creation of incentives that will encourage participants to care enough about the various aspects of climate change to want to make good policies. Examples of such incentives can include those that rely on economic arguments, peer pressure, civil society advocacy, and the enlightenment of decision makers. The idea of policy development implies that an area of policy has been defined and a weakness in it has been identified which can be corrected deliberately. New policy areas evolve through the advancing awareness that some things that had been previously ignored are in fact important, and evaluations can contribute to this process by helping stakeholders to look at systems in new ways. The Finnish country programme in Nepal illustrates this process, even though it previously featured in Chapter 2 by providing examples of high scores for relevance, impact and coherence, and will be mentioned again in Chapter 7 for its success in adapting to changing conditions of war and peace. Nevertheless, signs of subtle weakness appeared in the programme's medium score for connectedness in Chapter 2, which hinted at areas of policy in need of further development in ways that are likely to be relevant elsewhere (Box 6.27).

Box 6.27 Subtle weaknesses and policy development in Nepal

"**Water**. Finland's bilateral water supply projects have supported the creation of local systems that attract community support, participation, maintenance and cost recovery. Such systems are rightly seen as best practices, but there is the question of whether they give adequate attention to the management of water-bearing ecosystems in catchments (i.e. to sustainable supply rather than just delivery to users). Thus there is scope to consider building outwards from water user groups to embrace the safeguarding of water supplies in the context of better-managed forests, agro-forests, farmlands, rivers, aquifers and lakes in whole landscapes, including hydroelectricity potential, promoting 'payment for ecosystem services' (PES) arrangements to ensure that water-bearing ecosystems are properly managed on all scales, and the resolution of water conflicts between localities through equitable sharing of costs and benefits. In short, organising the delivery of potable water to users is only a fragment of the 'water sector' (albeit a necessary one, and hard to accomplish in poor, mountainous areas), and the concern is that this fragment, now absorbed into donor and government thinking and amplified through WASH for All [an incipient sector programme], may not be used as it should be: as an access point to the greater theme of holistic catchment management for sustainable development and climate change adaptation.

"**Forests**. A lesson from participatory forest projects in Nepal is that community forest user groups can be enabled through partnership with local institutions and the use of local facilitators to become robust and inclusive institutions that can attract pro-poor investment. Their net effect is to improve the condition of forests in hilly areas and help reduce rural poverty, and much community forestry work can also be expected to offer a model for future work on climate change adaptation. The potential is there, in principle, for the Multi-Stakeholder Forestry Programme (MSFP) to contribute strongly to the climate proofing of whole catchment systems, but there is some uncertainty over what the programme will actually do. It may be that an enhanced role for community forestry in promoting sustainable livelihoods and climate change adaptation could be achieved by Finland, either by guiding the MSFP in that direction if Finland participates, or in other ways if not, but in any case with greater synergy with the efforts of other donors and with government in a new, federal and decentralised context.

"**Climate change**. Nepal is exceedingly vulnerable to climate change, and its own vulnerability has important implications for the lives and livelihoods of hundreds of millions of people who live downstream in India. Finland in Nepal has taken small but strategic positions on managing relevant knowledge, both for climate change mitigation through investment in the forest sector, and for early warning of dangers resulting from climate change. Adaptation is all about resilience to change, reduction of risk and preparedness for disaster, and a case could be made for climate change adaptation to be a far more central feature of Finland's country programme. The Finnish-supported interventions to improve response capacities and manage early-warning data are tentative and small responses compared with the scale of the challenge. The WASH sector projects are just starting to contemplate the need for practical steps to climate-proof water catchments, but there is a need for a much more comprehensive approach based on integrated river basin management. This could be rooted in the water sector projects since they are oriented to the local level, precisely where adaptation activities are most needed. River basin institutions would also be appropriate focal points for early warning and response systems, governance and communication. While national/provincial policy frameworks and institutions are necessary, crucial adaptation action has to take place on the ground and Finland is well placed to help government take such action.

"**Comprehensive adaptation strategy.** If sustainable water supply is among the top development priorities at the district level, climate change adaptation is an increasing national priority, and the proper and participatory management of farm, forest and catchment ecosystems is the key to both, then it would make sense to consider how the various strengths of the current partnership can be used to create more complete solutions. One option would be to consolidate the water, forestry and climate change initiatives within community-based climate change adaptation. This would integrate the themes of resource and tenure mapping, multi-stakeholder forest management, catchment-friendly farming systems, and developing knowledge and procedures for preventing conflicts over water and for making PES arrangements between localities and countries. It would make use of the physical mapping work already accomplished with Finnish support, ... [with] districts being a particularly attractive scale at which to work because they are small enough to be accountably governed yet large enough to be meaningful catchment units, and most are bounded either by rivers or watersheds. Elements of this process could include:

- climate change adaptation planning by districts;
- participatory research on water, ecosystems and traditional knowledge to support resource management and conflict avoidance;
- catchment-wide knowledge-management arrangements; and
- catchment-wide governance of the climate-proofing process by an assembly of district representatives, supported by appropriate technical assistance."

Sources: Appendix 1 (33); Caldecott *et al.* (2012: 94–96).

References

ADB (2009) *The Economics of Climate Change in Southeast Asia: a Regional Review*. Asian Development Bank (Manila).

Asare, R. A., Kyei, A. & Mason, J. J. (2013) The Community Resource Management Area Mechanism: a Strategy to Manage African Forest Resources for REDD+. *Philosophical Transactions of the Royal Society B (Biological Sciences)* **368**(1625): 1–9. http://rstb.royalsocietypublishing.org/content/368/1625/20120311 (consulted 8 Apr 2016).

Bhatia, P. & Chugh, A. (2015) Role of Marine Bioprospecting Contracts in Developing Access and Benefit Sharing Mechanism for Marine Traditional Knowledge Holders in the Pharmaceutical Industry. *Global Ecology and Conservation* **3**: 176–187.

Caldecott, J. O. (1987) Medicine and the Fate of Tropical Forests. *British Medical Journal* **295**: 229–230.

Caldecott, J. O. (1988) A Variable Management System for Hill Forests in Sarawak. *Journal of Tropical Forest Science* **1**: 103–113.

Caldecott, J. O. (1997) *A Pilot Study on Bioindicators of Prevailing Micro-climates in Natural Rain Forests*. Development of Mapping and GIS for the Effective Planning, Management, Conservation and Sustainable Development of Forest Resources in Peninsular Malaysia (Project B7-5041/I/1992/10). EU Delegation & Forestry Department (Kuala Lumpur).

Caldecott, J. O. (2005) Lessons Learned and the Path Ahead. Pages 276–285 in *World Atlas of Great Apes and Their Conservation* (edited by J.O. Caldecott & L. Miles). California University Press (Berkeley).

Caldecott, J. O. & Indrawan, M. (2009) *Update of the Country Environmental Profile for Indonesia*. EU Delegation (Jakarta).

Caldecott, J. O. & McNally, R. (2008) *Final Report of the Mid-Term Review of the ASEAN Centre for Biodiversity*. EU Delegation (Manila).

Caldecott, J. O. & Pagett, R. (2006) *Regional Environmental Profile for Asia*. European Commission (Brussels).

Caldecott, J. O., Le, T. C., Muhtaman, D. R., Scutt, B. & Tan, E. E. (2002) *Final Report of the Mid-term Review of the ASEAN Regional Centre for Biodiversity Conservation*. EU Delegation (Manila).

Caldecott, J. O., Vouillot, A., Boonratana, R., Fernando, E., Setje-Eilers, U. & Fontaine, F. (2004) *Identification and Appraisal Mission for the Follow-Up to Project: ASEAN Regional Centre for Biodiversity Conservation (ARCBC) ALA/96/22 – Final Report*. EU Delegation (Manila).

Caldecott, J. O., van Praet, C. & Cai, M. (2009) *Final Report of the Mid-Term Evaluation of the EU-China Biodiversity Programme*. EU Delegation (Beijing).

Caldecott, J. O., Hawkes, M., Bajracharya, B. & Lounela, A. (2012) *Evaluation of the Country Programme between Finland and Nepal*. Evaluation report 2012:2. Ministry for Foreign Affairs (Helsinki).

Caldecott, J. O., Killian, B., Tommila, P., Rinne, P., Halonen, M. & Oja, L. (2013) *Scoping Mission for a Possible Renewable Natural Resource Economic Governance Programme in Tanzania: Final Report*. Norwegian Agency for Development Cooperation (Oslo).

Caldecott, J. O., Dornau, R., Gollan, J., Halonen, M., Saalismaa, N., Schilli, A., Simonett, O., Stuhlberger, C. & Tommila, P. (2014) *Technical Report on Effectiveness of the Swiss International Cooperation in Climate Change Mitigation and Adaptation Interventions 2000–2012*. Mandate of the Swiss Agency for Development and Cooperation (SDC) and the State Secretariat for Economic Affairs (SECO) (Bern).

CBD (undated) *Costa Rica: Payments for Environmental Services*. www.cbd.int/financial/pes/costarica-pestechnical.pdf (consulted 12 Sep 2016).

Child, B. (1996) Zimbabwe. Pages 122–138 in *Decentralization and Biodiversity Conservation* (edited by E. Lutz & J. Caldecott). The World Bank (Washington, DC).

Embrechts, J. V. (2008) *Monitoring Report: Regional Asia – ASEAN Centre For Biodiversity (MR-205569.02 – 03/06/08)*. EU Delegation (Manila).

FDA (2014) *FDA Approves Xtoro to Treat Swimmer's Ear*. United States Food & Drug Administration. www.fda.gov/NewsEvents/Newsroom/PressAnnouncements/ucm427274.htm (consulted 13 Sep 2016).

Forslund, J. (2007) *Wildlife Management on Communal Land in Namibia: An Economic Approach*. Uppsala, Sweden: Department of Economics at the Swedish University of Agricultural Sciences. http://ex-epsilon.slu.se:8080/archive/00001898/01/Johanna_

Forslund.pdf (consulted 16 Jan 2016).

Gardner, A. (2002) *Community Forestry in the Bamenda Highlands Region of Cameroon: a Partnership for Conservation*. Second International Workshop on Participatory Forestry in Africa. ftp://ftp.fao.org/docrep/FAO/006/y4807b/Y4807B12.pdf (consulted 16 Jan 2016).

Global Ocean Commission (2013) *Bioprospecting and Marine Genetic Resources in the High Seas*. Policy Options Paper 4, Global Ocean Commission (Oxford).

Hulme, D. & Murphree, M. W. (2001) *African Wildlife and Livelihoods: the Promise and Performance of Community Conservation*. James Currey (Oxford).

IUCN (2015) *Ghana Bridges Existing Local Forest Governance with Design of REDD+ Benefit Sharing*. www.iucn.org/about/work/programmes/forest/?22100/ghana-bridges-existing-local-forest-governance-with-design-of-redd-benefit-sharing (consulted 16 Apr 2016).

Jones, B. (2010) The evolution of Namibia's Communal Conservancies. Pages 106–120 in *Community Rights, Conservation and Contested Land: the Politics of Natural Resource Governance in Africa* (edited by F. Nelson). Earthscan (London).

Kiss, A. (editor) (1990) *Living With Wildlife: Wildlife Resource Management With Local Participation in Africa*. The World Bank (Washington, DC).

Living Earth (2008) *Community Resource Management Areas CREMAs (Ghana): Training Manual, Draft Constitution and Bye Laws*. Living Earth Foundation and Forestry Commission of Ghana (Accra).

Martin, R. B. (1986) *Communal Areas Management Programme for Indigenous Resources*. Department of National Parks and Wildlife Management (Harare).

Nelson, F. (2010) Introduction; the Politics of Natural Resource Governance in Africa. Pages 3–31 in: *Community Rights, Conservation and Contested Land: the Politics of Natural Resource Governance in Africa* (edited by F. Nelson). Earthscan (London).

Oyono, P. (2004) One Step Forward, Two Steps Back? Paradoxes of Natural Resources Management Decentralisation in Cameroon. *Journal of Modern African Studies* **42**(1): 91–111.

Reid, W. V., Laird, S. A., Meyer, C. A., Gomez, R., Sittenfeld, A., Janzen, D. H., Gollin, M. A. & Juma, C. (editors) (1993) *Biodiversity Prospecting: Using Genetic Resources for Sustainable Development*. World Resources Institute (Washington, DC), Instituto Nacional de Biodiversidad (Santo Domingo de Heredia, Costa Rica), Rainforest Alliance (New York), African Centre for Technology Studies (Nairobi).

Reliefweb (2016) *Mongolia: Dzud – Jan 2016*. http://reliefweb.int/disaster/cw-2016-000004-mng (consulted 12 Sep 2016).

Rodinson, D. & Felicisimo, D. Jr (2007) *Monitoring Report: Regional Asia – ASEAN Centre For Biodiversity (MR-20569.01 – 10/04/07)*. EU Delegation (Manila).

Ross, M. L. (2001) *Timber Booms and Institutional Breakdown in Southeast Asia*. Cambridge University Press (Cambridge).

Shah, T., Mavalankar, D., Azhar, G. S., Jaiswal, A. & Connolly, M. (2014) *Addressing Heat-Related Health Risks in Urban India: Ahmedabad's Heat Action Plan*. Climate and Development Knowledge Network. http://cdkn.org/wp-content/uploads/2014/05/Ahmedebad_Inside_Story_final_web-res1.pdf (consulted 29 Sep 2016).

TEEB (2010) *The Economics of Ecosystems and Biodiversity: Mainstreaming the Economics of Nature: A Synthesis of the Approach, Conclusions and Recommendations of TEEB*. United Nations Environment Programme (Nairobi).

ten Kate, K. & Laird, S. A. (1999) *The Commercial Use of Biodiversity: Access to Genetic Resources and Benefit-Sharing*. Earthscan (London).

Chapter 7

Evaluating partnerships

Partnership quality

In ordinary English, a 'partnership' is *an association between people who are involved in a common enterprise*, but in the aid world there is often the aspiration of a long-term relationship based on mutual aid between institutions, and hence an emphasis on equality. Partnership building describes the process by which such a relationship is cultivated, through mutual consent and recognition of mutual benefit. These nuances mean that an evaluator should pay attention to whether a relationship that is described by one side or another as a partnership actually is one. If people on both sides accept that a partnership exists then it probably does, but the units of exchange between partners can be diverse, and the terms of exchange are subject to interpretation. Thus, funding, service contracts, and university places might be offered by one side, while research permissions, policy change, and legislation may be returned by the other. The *existence* of a partnership can be confirmed using records of activities (visits, joint workshops, reports, etc.) which show that goods and services are being exchanged, and may provide information about the terms on which this happens. The *quality* of the partnership then depends partly on the frequency and content of these exchanges, but mainly on the extent to which participants are enthusiastic about them, which can be assessed using interviews. Both kinds of information can be used to assess performance against the 'partner satisfaction' criterion.

Diversity of partners

Aid donors may reduce their costs and political exposure by outsourcing responsibilities to partner institutions. More positively, partnerships offer access to the energies and local knowledge or contacts of charities and other civil society organisations, the synergies (and reduced risk exposure) available from jointly funding large enterprises that would otherwise not be affordable, and the benefits of diplomatic influence. The resulting partnerships in an aid programme can be very diverse, as shown from the Finnish (Box 7.1) and Swiss (Box 7.2) portfolios, and several options are available for the kind of donation (or, less often, soft loan) involved in funding them. One is to make a contribution to the core funding of an organisation, in which the grant may be 'untied' and hence almost entirely free of conditions, and another is to make an ear-marked grant, which can take one of several forms:

- *a contribution to a multi-donor thematic fund* (e.g. a sector-wide programme in a country, group of countries, or globally);
- *an agreement with an organisation to deliver particular services* in a sector, country, group of countries, or globally; or
- *a contribution to an organisation on condition* that it uses the money to carry out activities in a thematic area.

Box 7.1 Finnish contributions to organisations

Non-governmental organisations. Finnish NGOs undertake activities in all sectors of Finnish development cooperation, and are often highly effective. When giving grants to NGOs, the general approach of the Finnish Government is to respect their independence in choosing their methods, provided that there is no conflict with development policy. The three main MFA mechanisms for NGO funding are: (a) the NGO Partnership Scheme, which involves framework agreements with a small number of large NGOs; (b) project-based funding for Finnish NGOs; and (c) funding provided to three NGO foundations, which make small-scale grants to support activities in developing countries. Other MFA support to civil-society organisations (CSOs) in developing countries is provided through the Local Cooperation Fund managed by Finnish Embassies. In addition, the MFA finances various international NGOs, an NGO Service Centre (KEPA), which provides support services to Finnish NGOs, and the Finnish non-governmental development organisation platform (KEHYS RY), which supports NGOs on EU development policy issues.

Research institutions. Development research focuses on the circumstances of developing countries and features of the international aid system, and/or can involve research cooperation with individuals and institutions in developing countries, as well as capacity building and investments in developing country or international research institutions. In the evaluation sample, the main partners within Finland were nearly 120 research schools at 20 universities, plus research institutes that include those for food and agriculture (MTT), forestry (METLA), environment (SYKE) and health (THL). Multilateral partners in development research include the UN University and the UN Research Institute for Social Development, and within the sphere of the World Bank the Knowledge for Change Programme. Other prominent international partners were the CGIAR institutions (CIFOR, ICRAF, International Food Policy Institute, and International Livestock

Research Institute). Development research supported by the MFA in Finland is managed mainly by the Academy of Finland, while some is directly commissioned by the MFA. The Academy and MFA have intertwined Development Research Strategies that give primacy to the interests and independence of researchers, and neither the Academy-led nor the MFA-commissioned studies are strongly determined by specific MFA interests.

Multilateral actors. Finland acknowledges the importance of multilateral actors in Finnish bilateral and regional cooperation where it is necessary to rely on their expertise and work input in places and fields where Finland lacks its own foreign representation or resources. Otherwise, the main targets and channels of Finland's multilateral cooperation are the UN system (especially UNDP, Unicef, UNFPA and the WFP, but also UNEP, FAO, WHO, UNESCO and the ILO) and the development banks (i.e. the World Bank Group, African Development Bank, Inter-American Development Bank and International Fund for Agricultural Development). Finnish collaboration with these partners is based on joint programming and multi-year financing for thematic cooperation, as well as core funding. Thematic cooperation is guided by Finland's policy goal of supporting economically, socially and environmentally sustainable development and the perceived corresponding strengths of selected organisations. State fragility, peace-building and humanitarian assistance are areas of special cooperation between Finland and multilateral partners including the EU.

International organisations. The UN is funded partly by mandatory, General Assembly assessed contributions to its core costs and established organisations, based roughly on Gross National Income, and partly by voluntary contributions to its specialised agencies and programmes. Finland, with an economy less than a sixtieth the size of the USA's, makes mandatory contributions proportionate to the USA's 22 per cent of the UN core budget, and therefore has a small role from this point of view. However, Finland does make significant voluntary contributions to UN agencies – nearly US$150 million in 2009. Finland is well appreciated for such contributions, and exerts a kind of soft power within those UN agencies that it supports in this way.

Sources: Appendix 1 (10, 11, 14, 18, 20, 22); Caldecott *et al.* (2010).

Box 7.2 Swiss contributions to organisations

The survey identified 87 recipients among the following groups:

- *Research institutions*, mostly concerned with agricultural research, on farming systems, cultivar improvements, plant diseases, organic farming, and integrated pest management.

- *Thematic interest groups*, which focus on some particular subject or advance a particular cause, some being NGOs with a relatively narrow geographical and thematic focus (e.g. rural sustainable development, or land rights and natural resource management), others being regional in scope (e.g. addressing sustainable development in the Andes and Central Asian mountains, or ecological family farming and agrobiodiversity in West Africa), but most are of global scope and concern themselves with a great variety of issues (e.g. sustainable development learning, South-South linkage and cooperation, renewable energy promotion, and action on critical water issues).
- *Swiss NGOs*, with large contributions to Helvetas Swiss Intercooperation, SWISSAID, Bread for All, and the Aid Organisation of the Protestant Churches of Switzerland, reflecting the diverse and empowering work of the Swiss charitable sector.
- *Other multilateral institutions*, comprising the International Centre for Integrated Mountain Development, the Mekong River Commission, and the International Tropical Timber Organization.
- *The World Bank*, where contributions focused on:
  - The Forest Carbon Partnership Facility (see Chapters 1, 2, and 5).
  - The Programme on Forests, which was set up in 1997 to support analysis, innovation and knowledge-sharing with a view to promoting forest policies that would lead to improvements in areas ranging from livelihoods and financing, to illegal logging, biodiversity conservation and climate change.
  - The Commodity Risk Management Group, which is founded on the recognition of a strong link between risk management, financial stability, livelihoods of the poor, and development, and also an awareness of rising volatility in both weather patterns and commodity prices.
  - The Global Facility for Disaster Reduction and Recovery, which was established in 2006 as a forum of 41 countries and eight international organisations that seek to help developing countries reduce their vulnerability to natural hazards and adapt to climate change by mainstreaming disaster risk reduction and climate change adaptation in country development strategies.
  - The South East Europe and Caucasus Catastrophe Risk Insurance Facility, which aims to promote a regional catastrophe insurance market by providing access for homeowners and SMEs to affordable, but dependable and not subsidised, insurance cover against the risk of natural calamities such as earthquakes and floods, cover which has not been available in the commercial market.
  - Carbon Finance Assist, which is a multi-donor trust fund that was launched in mid-2005 to support capacity building and technical

assistance, initially focused on helping developing countries participate effectively in carbon markets, and to benefit from mechanisms established under the Kyoto Protocol (i.e. the Clean Development Mechanism and Joint Implementation), but later focusing on climate finance readiness, low emissions development, policy instruments and carbon pricing, and cities and climate change.
  - The Partnership for Market Readiness, which provides grant financing and technical assistance for capacity building and piloting of market-based tools for GHG emissions reduction.
  - The Climate Investment Funds, including the Clean Technology Fund and the Strategic Climate Fund.
- *Other international financial institutions*, comprising the Asian Development Bank and the Inter-American Development Bank (IADB), with all Swiss contributions targeting multi-donor financing facilities to do with the water and sanitation sector (e.g. the IADB's AquaFund).
- *The United Nations Development Programme (UNDP)*, with Swiss contributions comprising a large, long-term core grant, and a donation to the UNDP-managed Crisis Prevention and Recovery Thematic Trust Fund (a flexible funding mechanism designed for quick action following a natural disaster or violent conflict, or when a unique opportunity arises to reduce disaster risk or prevent conflict).
- *Other UN institutions*, including the United Nations Office for Disaster Risk Reduction, the Secretariat of the UNCCD, the Adaptation Fund of the UNFCCC, and the Global Climate Observing System of the World Meteorological Organization (WMO), which is designed to provide the comprehensive observations needed for monitoring the climate system, detecting and attributing climate change, assessing impacts of, and supporting adaptation to, climate variability and change, and research to improve understanding, modelling and prediction of the climate system.

Sources: Appendix 1 (45); Caldecott *et al.* (2014: 31–34).

These various kinds of support can blend into one another, partly because more-or-less conditional contributions may ultimately be intended to synergise, often at a higher, policy or global level, with thematic and site-specific projects or with clusters or series of projects. Thus a single contribution recorded in aid statistics might have both ear-marked and core funding components, or there may be separate contributions, each made on a different basis to the same organisation. Moreover, it is not possible for reasons of legal immunity to bind UN agencies to contracts, so memoranda of understanding (MoUs) are used instead to establish the purposes for which aid is given in these cases, while accepting that funds given to the UN cannot be independently audited, the MoU legally enforced, or those responsible for expenditure held legally responsible for it. In these circumstances, the boundary between earmarked funding and untied core funding is necessarily vague. It puts the UN agencies in a potentially difficult position, however, since a donor (or its public or parliament) may start to wonder how it can justify MoU-based rather than contract-based investment; to date, though, the political advantages of giving money to the UN have tended to outweigh the practical and legal issues involved in doing so.

Country partnerships

The examples in Boxes 7.1 and 7.2 omit what are often described as partnerships between a donor country and another country that for various reasons has been chosen as a focal recipient of the donor's aid programme. These bilateral relationships or country partnerships are often based on historical ties, which can usually be traced to former imperial or colonial experiences that have led to common languages, trade relationships or membership in post-colonial alliances such as the Commonwealth, or else more recent shared interests in ideology, religion or resource investment opportunities. In the case of Finland, which never had an empire and has had highly inconsistent relations with its major neighbours (i.e. Sweden, Russia and Germany) and their own ideological and military projects, the country's choice of partner countries is based on reasonably transparent criteria. These include:

- *the country's need for assistance* (e.g. its poverty level and the state of its environment);
- *support already received* (e.g. action by other donors and the level of development funding, joint programming processes, the role of multilateral actors and the EU);
- *the country's political situation and ownership* (e.g. the human rights situation, its own commitment to address development challenges, and the role of civil society);
- *the added value that can be offered to it by Finland*, and its administrative capacity for cooperation; and
- *the status of Finland's own policy priorities* in the country's development plans (e.g. the promotion of peace, security, and human rights so that poverty can be reduced sustainably).

In other words, the Finns logically consider what they themselves would like to achieve and choose partners accordingly. Once embarked upon, however, such country partnerships can become highly complex and variable over time, as the two sides adjust and readjust to one another in the light of all manner of events. Three examples shed light on the diversity of process and experience that resulted in 2002–2011 as these extremely complex systems responded to one another: in Nepal (Box 7.3), in Nicaragua (Box 7.4) and in Tanzania (Box 7.5).

Box 7.3 War and peace in Nepal, 2002–2011

Nepal is a mountainous, landlocked country lying between China and India. Its people are very diverse and number about 27 million, and there is a weak and recent tradition of democratic governance. Resentment of official corruption among the rural poor encouraged the Communist Party of Nepal-Maoist to launch a violent insurgency in February 1996, and the 'People's War' that followed took at least 13,000 lives before being resolved in November 2006. In April 2008 a Constituent Assembly was formed with the Maoists as the largest party, which dissolved the monarchy and declared Nepal to be a federal democratic republic. Since then, political instability and the absence of an agreed constitution and elected local government bodies have delayed the peace process and hindered overall development.

"Congenial working relations between the [Finnish] embassy and government seem to have been the norm in Nepal throughout the evaluation period. Interventions were consistently based on shared development objectives, modified at times by constructive negotiations around differences of emphasis. Although government would prefer greater progress on the Paris Declaration indicators, in particular more programme aid and government control over funding, this dissatisfaction is not with Finland alone but with the whole donor community. The evaluation team did not receive a single negative comment on any aspect of the country programme itself. Rather, observers were more interested in explaining its success, which is attributed (at least in 2006–2011) to the following factors:

- familiarity and close collaboration with the relatively small number of donors present in the country;
- an excellent team at the embassy and at the MFA regional/country desks;
- very good rapport between embassy personnel and the heads of mission representing other donors (especially the World Bank, DFID, ADB, SDC and several UN agencies);
- counterparts of consistently high quality in government (especially at the Ministry of Finance);
- a sense of optimism when the conflict came to an end, which faded to some extent but was sustained by the thought that peace itself was the most valuable 'peace dividend';
- helpful visits to Finland by the ministers of Finance and Education in 2008, the Prime Minister in 2009, and the Foreign Minister and Deputy Prime Minister in 2010;
- the fact that Finland represented the Czech Republic's EU presidency in 2009, which facilitated donor coordination and relations with government; and
- the fact that from May 2009 for 18 months there was a caretaker

government, during which senior officials remained in place, and unlike some donors Finland was willing to deal with the government despite its caretaker status.

"Considering the sectoral dimensions of the country programme:

- in the **forest sector**, the Ministry of Forests and Soil Conservation is positive over Finland's re-engagement;
- in the **education sector**, informants at the ministries of Finance and Education expressed appreciation of Finland's contributions (apparently regarding Finland as having the best education system in the world, and considering Finland's support for technical education and vocational training as particularly appropriate);
- in the **water, sanitation and hygiene sector**, Finnish inclinations towards maximum participation and inclusion align well with Nepalese policies, there is wide recognition and respect for Finland's work by other donors, and both the financial management and gender equity and social inclusion strategy that were developed by Finnish support have been replicated by other partners;
- in the **environment sector**, Finland's work in south-eastern Nepal has been commented on very positively;
- in the **peace-building, governance and human rights sector**, among other positive views it is thought that the extended dialogue, facilitated by International IDEA and the embassy, between the President of the Finnish Supreme Court and the Maoist leadership contributed to the relative independence of Nepal's Constitutional Court."

The country programme was therefore given a partner satisfaction score of 7.

Sources: Appendix 1 (33); quotation from Caldecott, Hawkes, Bajracharya & Lounela (2012: 81–82).

Box 7.4 Revolution and reaction in Nicaragua, 2002–2011

"Nicaragua's Pacific and Atlantic zones have very different histories [CIDCA, 1987], but since being united in the 1890s the whole country was dominated by an elite allied to US interests until a revolution led by the Sandinist National Liberation Front (*Frente Sandinista de Liberación Nacional*, FSLN) in 1979 [Walker & Wade, 2011]. Sandinist policies promoted a mixed economy, political pluralism, social programmes, and diplomatic and economic nonalignment, but its assertive nationalism and socialist features antagonised the USA, leading to the 'Contra War' of the

1980s. The Sandinists lost power in the 1990 election, and three ideologically conservative governments followed until the FSLN was re-elected in 2006. The government has since focused on restoring a Sandinist social programme while building a new alliance with Venezuela and the Bolivarian Alliance for the Peoples of Our America (*Alianza Bolivariana para los Pueblos de Nuestra América*, ALBA). The subsequent actions of the FSLN government were generally consistent with its previous term, in that it has kept its macroeconomic performance and policies acceptable to the international financial institutions, while doing little to antagonise the private sector and increasing the security of energy supplies to maintain business confidence and its own popularity. It also embarked on implementation of a pro-poor National Human Development Plan around which government activities were organised. A new feature at the time of the evaluation was an ambitious commitment to renewable energy, and a new context was the influence of Venezuela. Observers agreed that the government was being effective in at least the rural development, health and energy sectors, and there is some evidence of declining poverty and increasing equity. Even so, the assertive attitude of the government, and its apparent willingness to cut constitutional and electoral corners to stay in power, had led to deteriorating relations with traditional donors" (Caldecott, van Sluijs *et al.*, 2012: 19–20).

"The particular government partners of Finland in the health/gender and rural development sectors have expressed their satisfaction with the relationship, albeit with a preference for the common fund dimension and some reservations regarding the administrative load of individual projects. Other government entities have expressed strong dissatisfaction with aspects of the governance programme, and generally with the preference of traditional donors to leave Nicaragua rather than adapt to a new role through dialogue on terms set by government. Notwithstanding these positions, Nicaraguan government entities do show interest in an ongoing partnership based on dialogue and mutual respect. Moreover, there is likely to be much stronger satisfaction among non-governmental partners and local beneficiaries (often including local government) of the various projects and programmes, and this has probably been under-sampled in this evaluation. On the Finnish side of the partnership, opinions have become polarised over interpretations of the Sandinist programme and specific government actions" (*ibid.*: 81). The country programme was given a partner satisfaction score of 4, but it was thought that this would have been 6 before 2007, and that at the time of the evaluation it varied between 1 and 7 according to institution and stakeholder.

Finland later followed other donors in leaving Nicaragua to its own devices, a decision that the evaluation attributed at least in part to the flow of information between the embassy in Managua and MFA in

Helsinki. Here, "an important area where questions have arisen is in relation to descriptions of the economic and especially the political situation since 2007. This is because the wording of these reveals an unsympathetic view of the FSLN and its leadership, as well as scepticism over the policies, programmes and accomplishments of the government. Opinion is sharply divided among observers both at the embassy and the MFA over whether or not these accounts are fair and accurate. These reports are often the sole source of information for MFA personnel about Nicaragua, however, especially when not balanced by knowledgeable desk officers because of high rates of staff turnover. In these circumstances, anything other than scrupulous accuracy and balance carries with it the risk of distorting any decision that may be made concerning the future of the country programme. Thus, in a small embassy team the performance of each individual can be very influential, whether positively or negatively" (*ibid.*: 72).

Source: Appendix 1 (34).

Box 7.5 A one-party state and one-minister decision-making in Tanzania, 2002–2011

Tanzania has been governed by the Party of the Revolution (*Chama cha Mapinduzi*, CCM) since independence in 1961, despite multi-party general elections every five years since 1995. The image of CCM has been tarnished in recent years by a series of corruption scandals, and factionalism within CCM has also created a sense of instability in both government and parliament. Thus the evaluation covered a period that recorded the outcome of events within a partnership between Finland and a Tanzanian administration with a fading but still potent one-party mandate, in a country that had always emphasised dealing with donors as far as possible as a group, and during a time when the Finnish side was influenced by the decisions of a single strong-minded minister for development cooperation. This situation created several issues that were still working themselves out at the time of the evaluation. The net result was that some parts of the programme were performing well, because they were rooted in local knowledge, coordination with other donors, and strong analysis, and others not, because they had been created more suddenly and with little preparation. As a result, the evaluation noted that "there is the paradox that a programme of considerable achievement is being managed by people who have been demoralised to an exceptional degree by factors beyond their control" (Caldecott, Valjas *et al.*, 2012: 23).

These efforts to make distinctive investments outside the structure of government-donor relations could not immunise the Finnish country programme against contextual factors which were also confused. Thus the evaluation found that "considerable effort has gone into applying between donors and government the principles of alignment, harmonisation, aid effectiveness and coordination. Results, even if mixed, have been achieved in these jointly agreed areas, and globally Tanzania is one of the show cases of achievements of the [Paris Declaration] aid effectiveness agenda. Despite this, both government and donor representatives are strongly dissatisfied. The government side tends to blame donors, among other things, for intervening in its sovereign political matters at the technical level, for not sticking to the jointly agreed decision criteria (e.g. for general budget support), and for increasingly having recourse to modalities other than the programme-based ones. Donors, on the other hand, tend to claim among other things that government commitment to reforms has slowed down significantly, public financial management has deteriorated, and that government is not doing all it could to combat corruption. Trust has thereby been undermined, and there is a need to restore and improve dialogue. Some observers express a general satisfaction with the country programme and these are more abundant among Tanzanians than Finns. Where Tanzanians did express reservations, popular topics were the interventions in Mtwara, Lindi and Zanzibar, where Finns were described as being over-insistent on their own views and willing to abandon earlier cooperation initiatives without building on or learning from them. On the other hand, Tanzanians at the central government level compared Finland positively with other donors, blaming the latter for imposing too many conditions. The views of the embassy staff are consistent with general dissatisfaction on the donor side, but dissatisfaction was also detected in the MFA headquarters. Due to constraints on the Tanzanian side (e.g. in the management of the National Forest Programme and Local Government Reform Programme), the embassy and MFA have not been able to proceed as planned, and other modes of implementation have had to be found or planned disbursements redirected. Moreover, embassy and MFA officials are dissatisfied with the lack of transparency and collective decision making that led to the fragmentation of the country programme" (*ibid.*: 90). The country programme was given a partner satisfaction score of 3.

Source: Appendix 1 (35).

Sector-wide and basket funding

Forums and sector-wide programmes

Finding ways for different donor agencies to work together effectively on mutually supportive investments has long proven difficult. This is because each represents and responds to the needs of institutions in their own home countries, which may have different priorities and political aims, and certainly have different career structures, accounting rules, and bureaucratic procedures. In short, the officials within a group of donors all report in different directions and are accountable in different ways and to different people for their activities and expenditures, and this cannot be solved completely through good intentions and dialogue. The existence of forums such as multi-donor working groups and round-tables can at least be confirmed by an evaluation, and the extent of their activities documented from the records of their meetings. These forums often exist for extended periods, look good, but tend to yield few concrete results. Much better performance is achieved where thematic common funds or basket funds are established, to which governments each contribute and delegate spending decisions to aid officials who are responsible to each other for generating decisions rather than just meeting notes.

The last point can be generalised to sector-wide approach programmes (SWAp), in which all the factors that are active in a target economic sector are considered and addressed systemically, and also to regional programmes, in which connections, tensions, common themes, and opportunities for cross-border cooperation are the subjects of interest. Where there is shared, but individually accountable, responsibility for spending decisions, participants are likely to pay attention because they can see a direct relationship between their use of time (e.g. for meetings) and their receipt of financial resources (e.g. budgets) or other rewards (e.g. professional recognition). Combine this with a partner government that prioritises the target sector and much can be achieved by institutionalising the SWAp, while maintaining an ability to explore and test new ideas through innovative or strategic projects. Examples of SWAps founded on these principles in Nicaragua were assessed as having good strategic effectiveness, impact, sustainability, and coordination, as well as being potentially highly replicable to other circumstances where a similarly long-term, multi-donor commitment can be made (Box 7.6).

Box 7.6 The sector-wide approach in Nicaragua

"It seems that a particular combination of instruments in the health/gender and rural development sectors has been strategically effective and sustainable, with Finland and Nicaragua having designed and implemented distinctive and well-institutionalised sector-wide approach programmes (SWAps) that are robust to contextual change. The instruments concerned in this 'four-pronged' approach comprise the SWAp itself (aligned directly to government plans, priorities and institutions), sectoral budget support (to amplify and enable government participation in the SWAp), multi-donor basket funds (to ensure meaningful donor coordination based on shared spending decisions, and to support themes of importance to donors that also complement or supplement government activities) and specific projects to test, pilot or supplement actions in areas of particular importance to Finland or that make use of specific Finnish added value. A number of questions arise over this finding, concerning especially the degree of conscious design involved originally, the main sources of influence over the design (including the embassy, the MFA, other donors, the government, NGOs, and individuals associated with them), and the specific, time-bound (or process-bound) decision points involved. To explore these matters it was necessary to interview witnesses of the events of the period when the SWAps first came into being in 2004–2006. Based partly on the account of the then Rural Development Adviser (RDA), the following points can be made:

- An individual who had previously worked at the Nicaragua desk in Helsinki was appointed Ambassador to Nicaragua in September 2004.
- The RDA was transferred from MFA to the embassy in May 2005. She and the Ambassador had previously been working on the country programme for Nicaragua, and were equally familiar with the workings of the MFA, while both had been encouraged by the MFA to be innovative in developing a new SWAp. This coincided with the process of decentralising decision-making power to the Embassies that allowed a greater degree of independence in the formulation of programmes.
- At the time, the World Bank was taking the initiative on a rural development SWAp and, although it later retired, its presence contributed to there being enough potential financing available to justify a sector-wide approach, when combined with the active interest of Finland, Switzerland, Sweden and later Norway.
- It proved important that the actors should have a good knowledge of project management cycles and decision-making processes at their own headquarters, to allow the timing to be matched for the participation of different donors.
- It was also crucial to the long process of negotiation and adaptation involved that the embassy had available a particular consultant with long-term experience in Nicaragua, good negotiating skills and a willingness for the embassy to act as a low-profile catalyst.
- Thus the successful design and establishment of at least the rural development SWAp is attributable largely to the special combination of persons at the embassy and the trust and responsibility given to them

(by MFA to the Ambassador, by the Ambassador to the RDA, and by the RDA to the consultant)."

Sources: Appendix 1 (34); quotation from Caldecott, van Sluijs *et al.* (2012: 73–74).

Scaling and gearing through dialogue

Although SWAps and basket funds are vulnerable to the unanticipated withdrawal of contributors, as happened due to antagonisms of ideology and style between government and donors in Sandinist Nicaragua, the approach must be central to many investments in climate change because of the need to maximise collective performance across large parts of whole economic systems. This also applies to arrangements involving multiple countries in climate change mitigation and adaptation efforts, but here there are inherent challenges in what might be called 'scaling and gearing' between the global, inter-governmental level, and the level where resources are delivered to reduce emissions and manage ecosystems. A key feature of the REDD+ partnerships, for example, is the (painful but largely successful) search for such mechanisms (see Chapter 5). Within a SWAp and basket fund, or a REDD+ arrangement, there is a necessary role for countries whose representatives have a particular willingness, for cultural reasons, to invest effort in facilitating, supporting, and chairing meetings, and liaising among donors and governments to help solve problems at difficult times. These are roles in which the Finns (Box 7.7) and the Swiss (Box 7.8) tend to excel, and that are widely appreciated by other stakeholders.

Box 7.7 Finnish roles in multi-donor enterprises

"Lessons from success [in the Finnish country programmes in Nicaragua, Nepal and Tanzania] teach the value of: (a) making small but strategic and well-targeted contributions to multi-donor enterprises and combining them with the energy and competence needed to obtain respect and influence; (b) using well-formulated strategies for promoting participation, inclusion and equity in all contexts; (c) being willing to participate with government and donors in serious, long-term sectoral programmes; and (d) being able to preserve a complementary role for innovative and strategic technical experiments and partnerships that sometimes yield game-changing outcomes" (Caldecott, Hawkes, Bajracharya, van Sluijs, Aguilar, Valjas, Killian & Lounela, 2012: 1).

Sources: Appendix 1 (33–35).

Box 7.8 Swiss roles in the Adaptation Fund

The Adaptation Fund seems to be widely accepted among developing countries as a useful funding instrument for climate change adaptation, and has proved able to involve least-developed countries (LDCs) and small-island developing states (SIDS) fully, despite their capacity constraints. The fund was assessed as a vital contribution to the international climate finance architecture, and was given an adaptation effectiveness score of 6. Bearing in mind the persistent gap between what is promised in international climate negotiations and what is actually delivered, the Adaptation Fund can encourage other institutions to look for innovative financial solutions. The constructive and active role of the Swiss in the Fund's Board was praised specifically by several stakeholders in interviews and questionnaire replies, so a partner satisfaction score of 6 would also be appropriate.

Sources: Appendix 1 (45); Caldecott *et al.* (2014: 33, 367–370, project 7F-08274).

Core funding and partnership building

An important finding of the evaluation of the UNDP's strategic programme in 2013 was that a major reason for non-delivery of planned outputs was the under-resourcing of programming and projects due to the earmarked nature of most of its funds (UNDP Evaluation Office, 2013). This illustrates how funding only specific purposes, under contract or MoU, can starve an institution of the funds that it needs for all the other things that it has to do – the researching, thinking, discussing, networking, designing, and other tasks that are the life-blood of its activities, as well the investments in systems, growth, and profile that maintain its influence and effectiveness. For all organisations without a secure source of income, whether from government budgets, business activities, or a large subscriber base, untied core funding is the most valuable and best appreciated kind of funding. This is because it allows the organisation to build its capacity and programme activities in line with its aims, and particularly applies to NGOs at all geographic levels, where so much innovation and fine-grained attention to social and environmental detail occurs.

Core funding is often central to partnership building, and has a key role in partnerships with international organisations, regional entities, individual governments, research and educational institutions, and NGOs. An aim should be to avoid having to waste energy in justifying every grant by evaluating the performance of such organisations in each area where they work. Another should be to avoid the distortions that result when NGOs in particular become donor driven and distracted as they strive to bid for contracts. Conditional service-delivery contracts (or MoUs) look good to those that insist upon 'value for money' criteria, to the exclusion of subtler and longer-term aims, but tend to disempower the recipients, and eventually reduce their capacity to do good work. Rather, the aim in a true partnership should be to build the capacity of each partner, based on shared aims, reliable core funding, and training as needed, plus agreed-upon action priorities and additional project funding for work on those priorities. Thus, in conclusion, the process of partnership building should be based upon the following steps:

- the donor should first *determine its own priorities*;
- the donor should then *identify partners with an interest in similar priorities*;
- the donor should then offer each partner *untied core funding* at a level that will relieve it of fund-raising needs over an extended period;
- the donor should then, in the context of that relationship, work out with its partners what they both *agree is worth doing* in operational terms; and
- the donor should then offer *additional project financing* for undertaking those tasks.

References

Caldecott, J. O., Halonen, M., Sørensen, S. E., Dugersuren, S., Tommila, P. & Pathan, A. (2010) *Evaluation of Sustainability in Poverty Reduction: Synthesis*. Evaluation Report 2010: 4. Ministry for Foreign Affairs (Helsinki).

Caldecott, J. O., Hawkes, M., Bajracharya, B. & Lounela, A. (2012) *Evaluation of the Country Programme between Finland and Nepal*. Evaluation report 2012: 2. Ministry for Foreign Affairs (Helsinki).

Caldecott, J. O., Hawkes, M., Bajracharya, B., van Sluijs, F., Aguilar, B., Valjas, A., Killian, B. & Lounela, A. (2012) *Evaluation of the Country Programmes between Finland and Nepal, Nicaragua and Tanzania: Policy Brief*. Special edition 2012: 1. Ministry for Foreign Affairs (Helsinki).

Caldecott, J. O., Valjas, A., Killian, B. & Lounela, A. (2012) *Evaluation of the Country Programme between Finland and Tanzania*. Evaluation report 2012: 3. Ministry for Foreign Affairs (Helsinki).

Caldecott, J. O., van Sluijs, F., Aguilar, B. & Lounela, A. (2012) *Evaluation of the Country Programme between Finland and Nicaragua*. Evaluation report 2012: 1. Ministry for Foreign Affairs (Helsinki).

Caldecott, J. O., Dornau, R., Gollan, J., Halonen, M., Saalismaa, N., Schilli, A., Simonett, O., Stuhlberger, C. & Tommila, P. (2014) *Technical Report on Effectiveness of the Swiss International Cooperation in Climate Change Mitigation and Adaptation Interventions 2000–2012*. Mandate of the Swiss Agency for Development and Cooperation (SDC) and the State Secretariat for Economic Affairs (SECO) (Bern).

CIDCA (1987) *Ethnic Groups and the Nation State: the Case of the Atlantic Coast in Nicaragua*. Centro de Investigaciones y Documentación de la Costa Atlántica, Department of Social Anthropology, University of Stockholm (Stockholm).

UNDP Evaluation Office (2013) *Evaluation of the UNDP Strategic Plan 2008–2013*. United Nations Development Programme (New York).

Walker, T. W. & Wade, C. J. (2011) *Nicaragua: Living in the Shadow of the Eagle*. Westview (Boulder, Colorado).

Chapter 8

Evaluating transformations

Stability and sea change

In ordinary English, a 'transformation' is *a marked change in nature*, with the implication that an alteration to a deep structure has given rise to a new form or appearance on the surface. It is the idea of such a sea change that resonates, since it captures the notion that there are many things going on invisibly at depth within social systems, just as there are within people. They include the basic assumptions about relationships, priorities, and values that shape thoughts and behaviour at the social and individual level. These assumptions are usually unconsciously held, rarely challenged without resistance, and quite conservative, often being stable for decades for both people and societies. They may even be stable for centuries or millennia in the case of cultural and linguistic features of major human families, and in some cases they may be stable because they are features of the human species itself. Most do change over time, however, usually gradually but sometimes suddenly, either because something that had blocked change for a long time is removed, such as in the collapse of a longstanding political regime, or a new way of seeing the world is apprehended, such as in a religious conversion. In these cases, so many interlinked fundamentals are altered, that the net effect can only be a transformation in at least some aspects of society or personality. Minor, managed, and deliberate transformations are vital to aid investments, which seek to stimulate and guide a succession of miniature changes that are supposed to reveal themselves in an overall movement towards economic prosperity. In a larger sense, too, our response to climate change will have to involve a major transformation in our entire way of living if mitigation is to succeed. An understanding of how transformations occur, and how in practice they can be induced and directed, or predicted, detected and responded to, is therefore important to the evaluation of aid performance and the design of aid investments.

Historical latency

It may not be obvious to an evaluator who arrives in an unfamiliar place with ToR specific to the performance of an aid project, but they will be starting work amidst many potent ghosts. We have seen, for example, how making sense of an aid programme in Nicaragua or Nepal simply could not be done without awareness of local people's long experience of external political, economic, cultural, and military influences. Many countries took their modern form through imperial projects over the last several centuries, and the echoes of this experience are very persistent alongside many others. Moreover, parts of countries, at the landscape scale where aid projects often work, have their own unique circumstances that may differ profoundly from other parts of the same country. The practical implication is that for due diligence purposes an aid evaluator must take a strong interest in history (as well as ecology), and reading up on the subject and talking with local experts is an essential part of the job.

Two examples are provided to show how a transformation can be latent within historical circumstances, which an aid investment may activate, accidentally or on purpose, and which can then decisively influence performance, usually because it affects the enthusiasm with which local people respond to the initiative. In both cases there were distinctive historical experiences recorded continuously back to early in the Spanish imperial project, which shaped local aspirations and reactions into the present day. In the Chimalapas in México, remnant Amerindian peoples who had survived the conquest and its aftermath found a way to bribe the conquerors into recognising their ancestral domain, and took pains to renew the transaction whenever a change in governance occurred that was large enough to endanger the validity of the title deeds (Box 8.1). It remains to be seen whether the rewards of this persistence will survive the political and criminal turbulence of modern México, the effects of the North American Free Trade Agreement, infrastructure development, and climate change.

Meanwhile, in Samar island in the Philippines, history reveals an indigenous people left aside by events in the main centres of the archipelago, except for steady exploitation and occasional punishment for insurrection against outside rule (Box 8.2). In both the Chimalapas and Samar, projects designed to restore or reinforce a measure of local control of local resources inspired local people to attempt a transformation in their circumstances, only for both efforts to fall foul of non-local interests that misled or deterred donors into abandoning the principles on which their investments were based. Lest this be considered too negative by the reader, some potentially more positive outcomes are recorded elsewhere (e.g. on CBRM in Africa, in Chapter 6), and in any case up-dating of the Chimalapas and Samar cases may reveal something more hopeful, reminding us that history has no end while people and functioning ecosystems survive.

Box 8.1 Latent transformations – lessons of the Chimalapas in México

The Chimalapas comprise the contiguous communal lands of the two municipalities of Santa Maria Chimalapa (460,000 ha) and San Miguel Chimalapa (134,000 ha) in south-eastern México. They are located in the Isthmus of Tehuantepec, which separates the Gulf of Mexico to the north from the Pacific Ocean to the south; bounded to the north by Veracruz State, they overlap with Chiapas State to the east, and are otherwise within Oaxaca State. Together they contain about 450,000 ha of natural forest ecosystems, with scores of distinct types due to the effects of soils, drainage, elevation (approx. 400–2,250 m), rain shadow (approx. 400–5,000 mm annual rainfall), and cloud exposure. Both species richness and endemism rates are very high. The rain-trapping basin of the El Corte-Blanco river system in the heart of the Chimalapas rises from the rainforested lowlands to include about 60,000 ha of cloud forest in the surrounding mountains: the Sierra Atravesada to the south, Sierra la Ventana to the east, and Sierra Tres Picos to the north.

The Chimalapas are the remains of a larger area that was the ancestral domain of the Zoque people, and that they purchased from the Spanish administration in 1687. It is said that *Chimalapas* means 'gourds of gold' in the Zoque language, and that the gold involved was later used to build several palaces in Lima, Perú (which were lit at night by beeswax candles from Samar – see Box 8.2). The Mexican government confirmed title in 1850 after independence in 1821, and in 1883 after the liberal reforms of 1854–1874, but the area was reduced thereafter through a process of settlement and official mapping against a background of revolutionary and constitutional change. A federal decree in 1967 confirmed the dimensions of communal property in the Chimalapas at 594,000 ha, and the communal lands were defined again by the Oaxaca State delegation of the federal *Procuraduría Agraria* in 1998.

The story of the Chimalapas is that of a locally-owned frontier being settled or otherwise occupied and exploited by outsiders (cattle ranchers, loggers, narcotics traffickers, opportunistic land claimants, etc.). In response to this the communal land owners or *comuneros* (who are mostly peasant farmers or *campesinos*) have attempted to assert their ownership rights. An important tactic used by the Zoque as the original owners has been to recruit landless Indians as new settlers, to assimilate them as comuneros, and to locate them in unoccupied lands around the edge of the Chimalapas. The key process is thus the emergence of an enlarged ownership group with sufficient strength to hold on to the remains of the original (1687) Zoque land purchase.

A 'Campesino Ecological Reserve' was conceived as a federally recognised protected area, like a biosphere reserve but owned and managed by comuneros of the Chimalapas, comprising all who live there according to communal laws and by consent of the communal assemblies. The chief actor in pursuit of this strategy was a local NGO, Maderas del Pueblo, which had been leading efforts in the Chimalapas to resolve agrarian disputes and to promote a 'social fencing' process based on environmental education, participatory resource assessments, and community development planning. The enthusiasm of Maderas for this project is understandable in the context of a long-sighted strategy of survival by local people going back at least three centuries, which had resulted in legal recognition of the largest areas of municipal land tenure in the country. This attracted interest by the UK's country programme and a five-year project with DFID support began in 1995. A review of progress in early 1998 identified the following issues and lessons.

- The educational and participatory techniques used by Maderas are inherently empowering and may be indispensable for effective social fencing, but can be very slow especially at first when few experienced facilitators are available.
- Threats can build up much faster than social fencing, so while communities are empowered in one place destruction can be advancing in another, and even the most thoroughly empowered community can be overwhelmed by sudden threats backed by powerful economic, political, and military interests.
- Because educational empowerment is slow while threats are fast, there is a need to understand and neutralise threats to the area wherever they originate and not only within arbitrary geographical or institutional boundaries, while deliberately shielding the social fencing process until it becomes fully established.
- A key aim of investment is to establish a viable local system by closing down open-access use of resources, clarifying resource tenure, promoting social occupation of space, encouraging ecological understanding, promoting local self-identity, self-esteem and accountability, and encouraging the emergence of local rules governing the use of resources.
- This viable local system, however, must have good external relations, which means that another key aim is to promote local bargaining capacity and the ability to use outside laws, institutions, the media, etc., to defend local interests.
- A strategic vision of where the whole system should go is important, and this should be shared by donors and project implementers, and reinforced by regular and genuine communication.

A positive assessment of the ecological reserve in 1998 was insufficient to prevent the withdrawal of DFID support in 1999, as part of its general disengagement from México. This was driven partly by the country achieving middle-income status, as defined by national per-person GDP, and so becoming an unattractive site for aid investments. It may also have been encouraged by the 1994 Zapatista rebellion in Chiapas (Harvey, 1998), to which many campesinos of the Chimalapas were sympathetic due to their shared experience of landlessness and oppression. In these circumstances, British support for the restless poor in México may have been a political embarrassment in London.

Source: Caldecott *et al.* (1998).

Box 8.2 Latent transformations – lessons of Samar island in the Philippines

The island of Samar is the third-largest in the Philippines and the easternmost of the Visayan islands, with its eastern coastline facing the Pacific Ocean. The indigenous population is a distinct, Waray-speaking cultural group who were first contacted by Europeans in 1521. The island was annexed by Spain and ruled with the help of the Jesuits in 1596–1768, as a profitable source of products such as bees' wax for candles to light the great houses of the empire, and palm fibres with which to rig its ships. When the Jesuits were expelled from the empire in 1767 (the news, and the Franciscans to replace them, reached Samar the following year), the Franciscans took over the care of souls and trade in Samar until the Spanish were forced from the Philippines in 1898–1899 in the Spanish–American War. Indigenous rebellions by the people of Samar occurred in 1649–1650, 1869, 1884–1886, and 1894–1902, the latter coinciding with the beginning of American rule. Local resistance was crushed with great ferocity in 1902, and Samar was then governed by the United States against occasional armed resistance until mid-1942, when it was conquered by the Japanese and occupied by them until the Americans returned in late 1944. This troubled history of the Samareños continued, especially during martial law in 1972–1986 when the communist New People's Army was particularly active in Samar.

The net result is an island people who are used to preserving and occasionally asserting their cultural identity, but who are also used to being defeated when not being ignored. One consequence is a certain fatalism in outlook, combined with political marginalisation at the national level, and this has led to a pattern whereby young adult Samareños typically leave the island to live and work elsewhere. These factors together have ensured that Samar remains consistently poorer than other parts of the Philippines, with an economy largely dependent on remittances from abroad. Political weakness also helped to allow systematic exploitation of the island's forests for the benefit of non-Samareños during the 1970s and 1980s, which took away many of the islanders' natural resources and further contributed to local poverty.

Samar island receives an annual rainfall of 2,500–3,500 mm or more depending on location, with no distinct dry season but frequent typhoons.

The rugged interior was formerly covered by rainforest up to the maximum altitude of about 1,000 metres, growing on fragile soils derived from sedimentary, metamorphic and volcanic parent rocks, with large areas of *karst* (limestone) terrain. Despite the unsuitability of such ecosystems for high-intensity mechanised logging and conversion to agriculture, these have occurred in much of the island. Concerns over terrestrial wildland biodiversity in Samar are now focused on the survival of viable and representative samples of remaining natural habitats, including intact or regenerating rainforest at all altitudes and on different soil types. The island's lack of a central mountain range (unusual for the Philippines) means that much of the surviving natural forest grows at low elevation and are hence especially species-rich; they have been partly protected from logging and farming by the harsh topography and poor soils. A logging ban was declared in 1989 in response to public protest against floods and landslides, and survived despite attempts to re-activate timber licensing agreements. Much of the credit for preventing renewed logging in Samar island can be given to energetic local environmental NGOs, which have continued the Samareño tradition of idealistic militancy.

The idea of making better use of Samar island's biodiversity as a resource for the development of local society was explored from 1994 through the EU-funded WESAMAR Programme, which supported ten weeks of island-wide consultation and a two-day multisectoral workshop to design a Samar Island Biodiversity Project. This built on growing awareness of conservation issues in Samar, and on advocacy funded by the Foundation for the Philippine Environment through the Tandaya Foundation and other local groups. The outcome of this was a project design in March 1996 that enjoyed strong support locally and that had five main elements: (a) a sustainable financing mechanism in the form of a trust fund, (b) an island-wide biodiversity foundation, (c) a community-based conservation and environmental education process, (d) a biodiversity inventory process (with protection against 'gene piracy'), and (e) an area protection process. These elements were all to be fully accountable to the people of Samar island and to be implemented in a balanced and mutually-supportive way by them and on their behalf. This was to be done with the support of central government in the form of the Department of Environment and Natural Resources (DENR), which had approved the proposal.

Local buy-in to the March 1996 concept was unsurprising in the context of Samar's history, since for the first time an initiative was proposed for using Samar's resources for the joint benefit of Samareño and outside interests and with protections for the indigenous peoples of the island. A proposal to UNDP for the project was later developed by DENR, however, without local consultation, and without sharing the March 1996 paper; it excluded everything from the project except the area protection process, and placed that process entirely under non-Samareño direction. The March 1996 paper was quietly forgotten outside the island, UNDP and DENR proceeded according to their new plans, and Samareño stakeholders experienced a familiar disillusionment – as one put it, "As a people, we fight hard and then settle into passive resistance". From this story we can learn something about the power of hope and reason, which can leap up and burn brightly, and also the power of assumptions that disparage local participation and encourage exclusion of low-status stakeholders.

Sources: Schott (1964); Cruikshank (1985); Caldecott *et al.* (1996); Caldecott (1998a, 1998b).

Inducing transformations

With enough understanding of the historical and ecological circumstances of people dwelling in a landscape it should be possible to anticipate their likely enthusiasm for transformational change, and this should inform the design of any aid investment. Quite how this would shape the investment itself will depend on the appetite of donors and national authorities for sponsoring change, since a variety of interests, and both radical and conservative attitudes, will influence subsequent negotiations. Similar factors will be activated in response to changes that arise unexpectedly. For example, politicians may not object to a proposed environmental education programme at first, but may well do so when the awareness that it creates results in people actually protesting, in the media or on the streets, over land tenure, or over pollution or water issues caused by unregulated businesses, which may then escalate, fuelled by the memory of ancient injustice. All such consequences should have been thought through by project designers, but such perfect foresight is hardly routine. In any case, it is more usual for donor agencies (and politicians) to express a general aspiration for transformational change, without going into too much detail.

The Forest Investment Programme (FIP) falls under the Strategic Climate Fund, which was set up in 2008, is administered by the World Bank, and is financed by multiple donors (CIF, 2016). One of its aims is to initiate 'transformational change' in the LULUCF sector, by offering funding that forested developing countries can use to make reforms needed to allow REDD+ transactions to occur, such as on resource tenure, carbon rights, and benefit sharing. The idea is that, in order to undertake REDD+, a country's government would have to decide to change established patterns of 'who owns what, and who benefits from what' in the country's forests (i.e. by reforming forest tenure and governance arrangements), and close down industries and businesses that conflict with REDD+ objectives while encouraging others, all of these being very ambitious aims. It is the changing of tenure and governance that would amount to an alteration of the deep structure of the LULUCF sector, and a transformation should result.

This would be expensive, however, so the government concerned would need to be rewarded financially, diplomatically, and in market access and trade benefits, and provided with technical assistance, to facilitate the decision. Whether the source of the 'transformational change' in this process lies in the government or the FIP offer, or jointly, or elsewhere, is open to question. A government, after all, may make such decisions for various reasons of its own, such as in response to local public anger over flash-floods aggravated by deforestation, or to restore land rights that had been taken from indigenous people by a previous regime, or to clear away what it sees as obsolete land uses and claims in favour of new development strategies. In any case, it is not clear that adding the aim of inducing transformation to a list of donor objectives will produce any additional results, the more so as "the definition of transformation will depend on country circumstances; what is considered transformational in one country may be normal practice in another. Although it is a central principle of the FIP, transformation is often not fully understood by government agencies or other REDD+ stakeholders, especially in terms of its practical aspects" (CIF, 2013: 30; but see Box 8.3).

Box 8.3 Inducing transformations – lessons of the Forest Investment Programme

"Triggering transformational changes requires a programmatic approach that addresses the barriers to REDD+ in a systemic way. REDD+ initiatives need to support the implementation of national or sub-national strategies that shape policies and incentives, rather than isolated projects. Coordinating initiatives and organizing REDD+ funding streams at the national level is one of the most fundamental steps, and will allow for efficient and effective direction of resources and actions to REDD+ priority areas and issues. High-level political support is essential to sustain this approach. Some countries have opted for immediate implementation of small REDD+ and livelihood projects in order to show progress on the ground before a full consultation process has been carried out. However, these small-scale, isolated projects, while individually useful, are unlikely to contribute to transformation. The process of reaching consensus on a national REDD+ strategy and FIP investment plan should build a partnership among the stakeholders that will serve to support long-term REDD+ investments and transformational change in the pilot countries. This partnership or collaboration is itself a major product of the FIP. Lessons:

- What constitutes transformational change is country-specific and needs to be discussed and agreed on by REDD+ stakeholders during the FIP programming phase and monitored throughout the implementation of projects and programs agreed in endorsed investment plans.

- Achieving the FIP's mandate requires a programmatic approach that allows for targeted interventions that address the barriers to REDD+ in a systematic way. It must be understood that strong partnerships and transformational results will be achieved only over the long-term and with high-level political support.
- Building effective partnerships and consultation and collaboration processes should be viewed as a key outcome of the FIP and fundamental to transformational change."

Source: CIF (2013: 30–31).

On the other hand, in Chapter 2 it was mentioned that it can indeed be useful to think about inducing transformational change in designing, justifying, and assessing a project, in the sense of this being a goal that should be embedded within a theory of change to contribute to its sense of ambitious strategic purpose. After all, if an aid investment lacks the ambition to induce meaningful change, and its designers cannot explain how and why this will come about, why do it? Moreover, it is indeed possible to change how people think about things, through policies that validate, explanations that educate, and actions that mainstream, and this can affect the perception of each other's rights (e.g. by applying GESI strategies) and the value of natural ecosystems (e.g. by demonstrating PES arrangements or bioprospecting achievements). Once the point is made, practically, clearly and irrefutably, human–human and human–nature relationships may never be the same again, and a transformation will have begun.

Responding to transformations

Social systems occasionally transform themselves. This may occur at all sorts of levels and in various ways, from the outbreak of civil war to the agreement of peace, the triumph of a revolution or its reversal, the transition from a one-party state to a pluralistic democracy or back again, and, more directly relevant here, the sudden realisation that the health of national ecosystems is important to the population as a whole, rather than being of interest only to an elite as an exclusive source of private wealth. That these are complex and deep-rooted matters is shown by accounts of the back-story to the 2010 REDD+ break-through in Indonesia (Box 8.4), the role and history of INBio in Costa Rica (Box 8.5), Swiss interventions in the Amazon (Box 8.6) and Indonesia (Box 8.7), and the EU–China Biodiversity Programme (Box 8.8). The first makes the point that decisions may emerge only after a long sequence of contributing events; the second that intellectual leadership can exert profound, long-term effects; the third and fourth that a tiny amount of money spent in just the right way at just the right time can be almost infinitely more valuable than any amount of funds spent without adequate judgement; and the fifth shows that countries can quickly reach a new stage when new needs and priorities suddenly emerge or 'crystallise from the syrup of history'.

In no case is there the implication that history has ended and a stable endpoint achieved, for even a transformation can be reversed to some extent, although never entirely since people will remember and learn from what happened, and from what has been stifled or redirected. But in all cases, the idea is that donors should pay attention to the meaning of events among their country partners, and respond quickly and adaptively to opportunities that arise to support potentially game-changing developments. Another point is that formal evaluations, if not alert to the full significance of the course and direction of events, can yield erroneous conclusions by judging them, for example, on an instantaneous measure of efficiency. Thus, if an evaluator had asked in Costa Rica only 'how many bioprospecting contracts did INBio sign this year?', or in Indonesia 'how many trees have the Norwegians actually saved?', the strategic significance of the relevant programmes might be entirely lost.

Box 8.4 Back-story to the Indonesia–Norway REDD+ Partnership

Setting the stage. In 1961, most Indonesian forests were being legally managed under traditional rules by people who gave priority to long-rotation shifting cultivation and harvesting non-timber products for their own use from primary and old-growth secondary forests. A sharp change in the national regime and dominant ideology in the mid-1960s led to new ownership arrangements and new management aims for those forests. By 1971, they and the lands beneath them had been redefined as assets of the national development process, and control of the resulting 146 million hectare Forest Estate had been allocated in law to the Ministry of Forestry. Many of the extensive peatlands in Sumatra, Borneo, and New Guinea had been left out of this system, since at the time they were regarded as being of little use for forestry or agriculture. Over the following 15 years, with the technical advice of the World Bank, the UK, and others, the Forest Estate was divided into areas with different management aims: for conservation, conversion to other land uses, or log production. In the latter category, over 61 million hectares were licensed for logging by 1992–1993, with logs initially being exported whole, then as domestically-milled plywood, then as pulp and paper. It soon became more attractive to pulp logged-over forests than to wait for them to recover, and once the forests were pulped, the land became available for plantations. Presiding over this transformation of forest lands, the Ministry of Forestry became a powerful and secretive institution financed by the sale of logging and plantation permits for land within the Forest Estate.

The 1998 transition. These orderly arrangements became deeply entrenched prior to the next change of regime in 1998, which was followed by four years of chaotic governance and hasty decentralisation. District authorities took the opportunity to extend their role in authorising forest exploitation and conversion, which they did by issuing logging and plantation licences to their constituents. Local officials and freelance actors then also participated in a rapid expansion of illegal logging and forest clearance throughout the lowlands, including the conservation areas. These factors contributed to a deforestation rate that peaked at nearly two million hectares annually in 2002–2003. Continuing deforestation at close to a million hectares annually thereafter remained largely driven by conversion of forest and peatland to plantations, particularly of oil palms (*Elaeis guineensis*) and pulpwood trees (especially species of *Acacia* and *Eucalyptus*). Such activities can be extremely profitable, and the interest groups they create represent formidable opposition to any change from business as usual.

Action and reaction in the 2000s. In its efforts to retrieve a measure of control over the Forest Estate, after the events of 1998–2002, the Ministry of Forestry collaborated with the presidency and the rest of government in acting against unauthorised logging, with international actors in developing a national timber legality assurance system (SVLK), and with local government through participation in their cross-sectoral development planning procedures. It also pushed back decentralisation through a series of government regulations in 2004–2007, on forest planning, protection, and management, on monitoring the conduct of local governments, and on dividing responsibilities between the centre and the regions. The Ministry thus managed to preserve a key role for itself in governing the Forest Estate, now including very large areas that had been wholly deforested. It has since been adept at finding ways to profit from the lands under its jurisdiction, mainly from plantations but also including reforestation, nature-oriented tourism, and the sale of conservation concessions to environmental charities. It was also quick to recognise the financial potential of carbon conservation as a new forest management aim.

Sources: Appendix 1 (3, 32, 36, 47); Jepson *et al.* (2001); Ross (2001); Ribot *et al.* (2006); Caldecott & Indrawan (2009).

Box 8.5 The role and history of INBio in Costa Rica

Planning. Most of Costa Rica's almost five million people live in the Valle Central around the capital San José, where the economy is based on farming, light industry and tourism although elsewhere in the country plantations and cattle ranches are widespread. Most of the country's 51,000 sq. km land area is privately owned and, since government has little power to regulate use of private land, its owners were free to invest in forest clearance. During the 1970s and 1980s the country had one of the world's highest rates of deforestation, with almost 4 per cent of forest land cleared annually in 1981–1985. This convinced local conservationists that virtually all private lands were destined to be cleared of natural ecosystems, and that if most of the country's wild species were to survive, they would have to do so in nature reserves under public ownership. During the 1980s, some of these actors moved into positions where they could influence government, and their lobbying led to the presidential appointment of a National Biodiversity Planning Commission in 1989, the role of which was to design a comprehensive strategy for conservation. It started its work on the assumptions that biodiversity was economically valuable, and should therefore be preserved and used for public benefit, and that forest protection would in the end be unsuccessful unless the people living around each protected area were willing to help protect it. The Commission proposed new legislation to consolidate a National System of Conservation Areas (SINAC), with all the units being managed locally and for local benefit. To help pay for this, the Commission recommended a national biodiversity inventory, to find out exactly what made up Costa Rica's biological richness and what it might be used for, and also the creation of a National Biodiversity Institute (INBio) to manage the inventory.

Bioprospecting. By late 1993, INBio possessed a well-established and rapidly-growing infrastructure in Heredia, near San José, and was running 28 biodiversity stations in all ecological zones of the country. These were the field sites of the national biodiversity inventory, which was then yielding specimens for processing by INBio at a rate of about 900,000 each year. The inventory was designed partly with a view to the development of commercial products through bioprospecting, so an important aim was for Costa Rica to obtain a share of any resulting revenues, in order to help pay for national conservation efforts. INBio therefore sought to ensure that all biodiversity research occurred under legal contracts, which guaranteed local returns on commercial products. All through the 1990s, therefore, INBio forged bioprospecting partnerships with a series of pharmaceutical and biotechnology companies, and in the process learned how to formulate workable contracts in which the option existed to take up-front payments in kind, as technology transfers and training rather than in cash, while retaining a stake in the long-term success of the bioprospecting initiative through the contracting of royalties on any profits that may later be made. Between 1991 and 1999, INBio's bioprospecting agreements contributed more than US$390,000 to the Ministry of Environment and Energy, US$710,000 to conservation areas, US$710,000 to public universities, and US$740,000 to other groups at INBio, a total of US$2.55 million.

Continuation. Details of bioprospecting successes in Costa Rica since 2000 are sketchy, as they are in other countries and territories known to be actively developing bioprospecting partnerships, including the Malaysian states of Sarawak and Sabah, and Singapore (with other activity in South Africa, Thailand, Indonesia and the Philippines). This is to be expected, given that commercial secrecy arrangements are integral to bioprospecting agreements, and some jurisdictions regard the whole subject as subject to official secrecy rules. But this lack of information has sometimes made it hard to convince countries to invest political and financial capital in bio-diversity supported by inventory and bioprospecting. Several factors also eventually led to reduced interest in bioprospecting by large companies in tropical environments. These included increasing bureaucracy in some places, driven by a fear of biopiracy and the need to invent laws and procedures in each place with little specific reference to past experience (including Costa Rica's), the slow progress, long lead time and high costs involved in bringing new products to market, and the general rise of financialised capitalism that values quick profits over strategic investments. Although bioprospecting continues, as noted in Chapter 6, the sense that it could single-handedly motivate and finance the end of biodiversity loss and mass extinction has passed. The idea of purposive inventory of unknown wild species nevertheless contributed hugely to the inspiration of a generation of conservationists, and certainly remains valid and transformational. But back in its homeland, INBio had attempted to diversify its own finance flows by investing in an environmental education theme park, but was crippled by interest payments and was being fully reorganised at the time of writing.

Sources: Caldecott (1996); ten Kate & Laird (1999); Conniff (2012); Pennisi (2013); Fonseca (2015).

Box 8.6 A Swiss intervention in the Brazilian Amazon

The Amazon Fund is the largest dedicated fund supporting efforts to reduce emissions from deforestation and degradation in the Amazon, and is managed by the Brazilian Development Bank with US$1 billion in funding from the Government of Norway, and technical assistance from Germany. The 16th UNFCCC CoP (2010) agreed that REDD+ investments must be adequately safeguarded to avoid injuring the land and other rights of indigenous people, and to encourage and enable their participation in designing and implementing such investments and in benefits arising from them. With the establishment of the Amazon Fund, it became vital to ensure the involvement of the many indigenous peoples of the Amazon

Basin. Obtaining this participation across such a large area was hard but necessary, and the project provided a US$1 million grant to support indigenous peoples in networked dialogue across the Amazon to allow the expression of their independent opinions on REDD+ and its implementation. This contributed to unlocking the potential of the Amazon Fund, so it was given a mitigation effectiveness score of 7.

Sources: Appendix 1 (45); Caldecott *et al.* (2014: 18–19, project 7F-08110).

Box 8.7 A Swiss intervention in Indonesia

A grant via the World Bank to the Indonesian REDD+ Task Force supported its Strategy and Financing Instrument working groups at a critical time, and also financed inputs on the design of the Indonesian REDD+ Financing Institution. The contribution was described by the evaluators as an example of a relatively small grant contributing to disproportionate impacts, because "by providing less than US$1 million in 2011–2012 to support the work of the Presidential Task Force on REDD+ in Indonesia [it] helped create a 2011–2015 moratorium on new logging and plantation concessions, which in August 2013 was estimated by the World Bank to offer benefits worth at least US$500 million". It was therefore given a mitigation effectiveness score of 7.

Sources: Appendix 1 (45); Caldecott *et al.* (2014: 32, project UR-00544).

Box 8.8 The EU–China Biodiversity Programme

"One field project stakeholder made the point that peasant farmers in China are much more interested in education than in money. This might also stand for the attitude of government towards foreign-assisted projects: that they are much more interested in new ideas than in the funds themselves. It is clear that money is not really a constraint in China, as a result of economic growth, stimulus budgeting and the availability of the ecological compensation and forest restoration funds. Knowledge of how to spend money wisely, however, is much more limited, and new ideas are welcome. Thus the key to achieving sustainable results is to introduce new priorities and new ways of thinking and working. This is an unlocked door in China, where government has recently called for 'science-driven development planning', and where the staff even of provincial forestry bureaux are writing monthly essays on how to apply science to development. Throughout the mid-term evaluation it was clear that Chinese officials at all levels were acutely interested in learning new things. This bodes well for the sustainability of [the programme's] impacts at a policy level, and suggests potential for even greater influence if the strategic processes now under way (e.g. with the China Council for International Cooperation on Environment and Development) are successful, and if the knowledge resources generated by the [programme's] field projects can be harnessed effectively. Government has learned the value of clean air and water, and is quickly learning about the need for well-functioning ecosystems and biodiversity. Thus the [programme] has an historic opportunity to introduce new and better policies which will influence events all over China for decades to come" (Caldecott *et al.*, 2009: 28–29).

Source: Appendix 1 (9).

References

Caldecott, J. O. (1996) *Designing Conservation Projects*. Cambridge University Press (Cambridge).

Caldecott, J. O. (1998a) *Report on a Visit to the Philippines for UNDP to Support the Samar Island Biodiversity Project*. United Nations Development Programme (Manila).

Caldecott, J. O. (1998b) *Letter of Resignation as Samar Island Biodiversity Project Development Expert (12th November 1998)*. Addressed to United Nations Development Programme (Manila).

Caldecott, J. O. & Indrawan, M. (2009) *Update of the Country Environmental Profile for Indonesia*. EU Delegation (Jakarta).

Caldecott, J. O., Piczon, E. C., Aliposa, J. S. & de la Cruz, M. T. (1996) *Samar Island Biodiversity Project Concept Paper*. Western Samar Agricultural Resources Development Project (Catbalogan).

Caldecott, J. O., Boege, E. & Musi, J. (1998) *Conservation in the Chimalapas, Isthmus of Tehuantepec, México*. Department for International Development (London).

Caldecott, J. O., van Praet, C. & Cai, M. (2009) *Final Report of the Mid-Term Evaluation of the EU–China Biodiversity Programme*. EU Delegation (Beijing).

Caldecott, J. O., Dornau, R., Gollan, J., Halonen, M., Saalismaa, N., Schilli, A., Simonett, O., Stuhlberger, C. & Tommila, P. (2014) *Technical Report on Effectiveness of the Swiss International Cooperation in Climate Change Mitigation and Adaptation Interventions 2000–2012*. Mandate of the Swiss Agency for Development and Cooperation (SDC) and the State Secretariat for Economic Affairs (SECO) (Bern).

CIF (2013) *FIP Learning Product: REDD+ Stakeholder Collaboration at the Country Level*. Climate Investment Funds Administrative Unit for the European Bank for Reconstruction and Development, African Development Bank, Asian Development Bank, Inter-American Development Bank, and the World Bank Group (Washington, DC).

CIF (2016) *Climate Funds Update: the Funds*. www.climatefundsupdate.org/data/the-funds-v2 (consulted 13 Jan 2016).

Conniff, R. (2012) A Bitter Pill. *Conservation* (University of Washington). http://conservationmagazine.org/2012/03/a-bitter-pill (consulted 19 Jan 2016).

Cruikshank, B. (1985) *Samar: 1768–1898*. Historical Conservation Society (Manila).
Fonseca, P. (2015) Costa Rica's INBio Facing Government Bailout. *Scientific American*. www.scientificamerican.com/article/a-major-center-of-biodiversity-research-crumbles/ (consulted 19 Jan 2016).
Harvey, N. (1998) *The Chiapas Rebellion: the Struggle for Land and Democracy*. Duke University Press (Durham, North Carolina).
Jepson, P., Jarvie, J., MacKinnon, K. & Monk, K. A. (2001) The End of Indonesia's Lowland Forests? *Science* **292**: 859–861.
Pennisi, E. (2013) Costa Rica's INBio Facing Government Bailout. *Science* (American Association for the Advancement of Science). www.sciencemag.org/news/2013/04/costa-ricas-inbio-facing-government-bailout (consulted 19 Jan 2016).
Ribot, C. J., Agrawal, A. & Larson, A. M. (2006) Recentralizing While Decentralizing: How National Governments Reappropriate Forest Resources. *World Development* **34**(11): 1864–1886.
Ross, M. L. (2001) *Timber Booms and Institutional Breakdown in Southeast Asia*. Cambridge University Press (Cambridge).
Schott, J. L. (1964) *The Ordeal of Samar*. Bobbs-Merrill (Indianapolis).
ten Kate, K. & Laird, S. A. (1999) *The Commercial Use of Biodiversity: Access to Genetic Resources and Benefit-Sharing*. Earthscan (London).

Chapter 9

Conclusions

Improving aid performance

Statements of aggregate performance

Naïve observers may instinctively seek to classify aid investments into the categories of 'dud' or 'effective'. At first glance, this conflicts with the desire among aid professionals for a nuanced appreciation of the design and performance of investments, but, provided complexity and realism are not neglected, it seems fair to adopt performance thresholds for the purposes of reporting results to public and parliament. This approach was demonstrated in Chapter 2 (Table 2.1), where four of six investments were considered to be potentially 'effective', and one or two of them were seen as potential 'duds', based on their mean scores across all core performance and design quality criteria. This approach can draw attention to the weaker investments in a portfolio, which can then be corrected or cancelled, and this will have the effect over time of raising the mean score of the portfolio as a whole. A version of this process, driven by the evolution of policy and attention, is visible in the Finnish and Swiss portfolio analyses (see below), but first there are two other factors to consider which can lead to high performance, respectively in the short and long term.

The rewards of 'cutting diamonds'

A few of the cases mentioned in this book draw attention to how tiny investments made in the right way at just the right time can achieve dramatic change, generally because the people designing them appreciated and took advantage of circumstances that may have been created by others, or by history, or by accident. They include the small Swiss interventions to facilitate potentially game-changing REDD+ initiatives in the Brazilian Amazon (Box 8.6) and Indonesia (Box 8.7). In principle, they also include investments that might have achieved much more than they did, but were ultimately thwarted by other factors. They include the cases of the National Biodiversity Institute in Costa Rica (Box 8.5), the role of sustainable charcoal in the management of Village Land Forest Reserves in Tanzania (Box 4.6), the Campesino Ecological Reserve in México (Box 8.1), and the Samar Island Biodiversity Project in the Philippines (Box 8.2).

It can also be important to be able to see the significance and potential gains to be obtained by doing something fairly non-obvious. Examples include combining multiple facilities to promote green investments in different ways in Vietnam (Box 4.18), exporting used trams to Romania and Ukraine (Box 4.14), off-setting the impacts of liming soils in Kenya (Box 5.8), and rehabilitating ancient *kahriz* water systems in Azerbaijan (Box 6.21). These all required a sensitivity to the true complexity, historical depth, and inter-linked nature of the real world. It is not yet clear whether this can be trained for specifically, or is a side effect of the process of paying attention to the core evaluation criteria. It is, however, a useful skill that is worth seeking, like the talent for understanding the structure of a gemstone so that it can be cut with precise strokes to show its best facets.

The rewards of consistent, strategic effort

Inspiration, intuition, and insight are not common in the cases considered in this book (though they might be more so than official reports suggest), and problems that have arisen from humanity's tenure of the biosphere are more often overcome from many years of consistent, strategic effort. Thus, for example:

- *REDD+ arrangements* once seemed to be impossibly difficult to make, but with the Forest Carbon Partnership Facility (Box 5.4) and the Indonesia–Norway REDD+ Partnership (Boxes 5.3 and 8.4), we now know that it is not impossible to do, just very hard;
- *the rate of natural forest loss in Costa Rica* once seemed terminal, but now with PES arrangements in place after being pieced together from previous and new laws (Box 6.11), deforestation there has been halted and even reversed;
- *the biodiversity crisis in South-east Asia*, which endangers some 40 per cent of all the species on Earth, once seemed of no interest at all to the mainstream leadership of any ASEAN country, but now with the investments of the EU and Germany in the ASEAN Centre for Biodiversity (Boxes 6.6–6.9), this is gradually changing; and
- *illegal logging in Indonesia*, and the corruption, smuggling and land clearances linked to it, once seemed insurmountable, but now with a combination of shared strategic effort by the Indonesian, UK and several other governments and the EU, *and* NGOs, *and* the general public, *and* the corporate sector (Boxes 5.9–5.11), progress is quickly being made.

The same point about the value of slow, steady effort could be made of the progress found within the bilateral aid portfolios considered here. The Finnish programme was not focused on climate change, but with experience and awareness of emerging environmental and other challenges it showed increasing attention to them (Box 9.1). The Swiss efforts were particularly relevant to the climate change response, so the moderate-to-strong overall score across 423 investments was encouraging, as was the higher average score from 2007 onwards for both mitigation and adaptation, and especially for adaptation (Box 9.2). By all these means, the 'baby steps' mentioned in Chapters 1–2 have often been taken, and we are now ready to move up a gear.

Box 9.1 Finnish aid, 2008–2010: diversity, strengths, and challenges

"The Finnish aid programme is diverse, with numerous modalities that in principle offer officials and other stakeholders a range of options for tailoring different combinations of activity with different combinations of need and opportunity. The programme is also diverse in its choice of targets, themes, sectors and locations. This diversity is both a strength and a weakness: a strength because the world is complex and adaptive flexibility and experimentation are therefore essential; but a weakness because it can fragment attention and resources into packets that are each too weak to meaningfully address all of Finland's development priorities. An impression is that as new priorities have been identified they have been *added* to the cooperation mix, rather than always *substituted* for previous priorities (the CCTs being a good example).

"Even in the current situation, though, we have detected a number of strengths in the delivery of this programme, as well as some weaknesses that offer opportunities for further improvements. We say 'further' here because there are signs that the effectiveness of aid delivery is already improving, and that changes are underway that will accelerate this by correcting some of the weaknesses that were identified by retrospective studies. Meanwhile, however, the 'outside' world – meaning the world beyond Finland's orderly and peaceful prosperity, but within which Finland nevertheless exists – is changing rapidly and bringing new challenges. Some of the signs of improvements in aid delivery that have been noted are responses to major emergent challenges [i.e. climate change, water security, food security, social problems and state fragility, ecosystem degradation and biodiversity loss], and are building enhanced preparedness for achieving poverty reduction objectives in a changing world."

Sources: Appendix 1 (10–31); Caldecott *et al.* (2010: 106–107).

Box 9.2 Swiss climate change investments, 2000–2012: hallmarks of effectiveness

Overall effectiveness was moderate to strong. This applied equally to mitigation, adaptation, and the strengthening of enabling frameworks for climate action in developing countries. It implies that Swiss public funds allocated to these kinds of action have been used effectively, and have produced results that support low-carbon and climate-resilient development in partner countries.

Effectiveness has improved over time. This is shown by comparing the 2000–2006 and 2007–2012 parts of the portfolio. Although exceptions were found, the trend holds overall for both adaptation and mitigation although it is more marked for adaptation. It implies institutional learning and may reflect the increasing policy priority given to climate change. The faster rate of improvement for adaptation suggests a steeper learning curve as adaptation has moved up the policy agenda, and the trend is expected to continue with increased international commitments to promoting adaptation.

Considering climate change in design improves effectiveness. There was a positive correlation between the extent to which climate change was considered in the design of investments and the later strength of their climate effectiveness. Comparing the 2000–2006 and 2007–2012 parts of the portfolio, there was also a clear increase over time in the extent to which climate change was considered in design, presumably responding to increased policy priority and the introduction of the Rio Climate Markers over the same period.

The most effective mitigation projects were:

- those that minimised new GHG emissions by avoiding new construction, or by re-using facilities, equipment and materials in which GHG emission costs had already been incurred or potential emissions embedded (e.g. Boxes 4.1, 4.3, 4.10, 4.14);
- those that built strategic capacity to manage and share knowledge and to leverage change by exploiting new opportunities created by mitigation investments (e.g. Boxes 4.12, 4.15, 5.4, 6.25); and
- those that joined together complementary mitigation initiatives in a structured way to promote synergy and long-term change (e.g. Boxes 4.8, 4.15–4.18).

The most effective adaptation projects were:

- those that promoted the reduction or sharing of disaster risk (e.g. Boxes 6.17–6.19);

- those that promoted the management and sharing of knowledge on vulnerabilities and adaptation solutions (e.g. Boxes 6.23–6.26); and
- those that rewarded the sustainable management of ecosystems and ecosystem goods and services through local empowerment and financial rewards (e.g. Boxes 6.2, 6.20, 6.21).

The projects that were most effective at both adaptation and mitigation were:

- those that leveraged investments in the conservation of forest, grassland and soil ecosystems (e.g. Boxes 5.4, 6.11, 6.13, 8.6, 8.7);
- those that promoted the flow of knowledge into policy development processes (e.g. Boxes 6.12, 6.17, 6.25, 6.26); and
- those that offered core funding to allow selected, effective, research, charitable, financial and UN institutions to improve the programming of their climate-related activities (e.g. Boxes 4.8, 5.4, 7.2, 8.7).

The projects with particularly weak scores for adaptation and mitigation were:

- those with poor awareness and communication of climate impacts (e.g. Box 2.6, 2.24, 2.26);
- those that showed a lack of attention to social and ecological aspects of water supply (e.g. Box 6.19);
- those that made an inappropriate choice of biotrade targets (e.g. Box 6.4); and
- a small number (not described here) that had a focus irrelevant to climate change, which therefore should not have been included in the portfolio in the first place.

Scarcity of quantitative data. Although there was ample qualitative evidence to support these findings, quantitative data were very scarce:

- **for mitigation**, because few data on emissions were collected and baselines were seldom defined (which is a weakness that requires correction in aid portfolios generally);
- **for adaptation**, because there are no agreed international standards for quantifying processes that maintain diversity and build resilience (which it may never be possible or desirable to correct, although rational and evidence-based assessments of adaptation investments are certainly needed).

Sources: Appendix 1 (37–46); Caldecott *et al.* (2014: iii–v, 50–53).

Improving aid evaluation

Sensitisation to options

Chapter 1 explained that this book has two connected aims:

- to make a case for the universal use of an extended version of the DAC evaluation criteria, with common definitions and scoring methods; and
- to provide explanations and examples of how to use the eight core evaluation criteria, and other key considerations relevant to partnerships, policies, and investment designs.

This approach uses multiple points of view to build up the fullest-possible appreciation of each aid investment as a complex, multi-faceted system in its own historical, social, and ecological context. As this approach was described and illustrated, the reader was to be sensitised to the rationale and options for various kinds of investment aimed at climate change mitigation using technology and ecology, and adaptation by preserving resilience and diversity, as well as to some emerging priorities, policy issues, and opportunities. The next section considers how to train those who evaluate aid investments to apply the lessons more effectively in their work. Since it is clear from Box 9.2 that paying attention to climate change in the design of investments helps determine their eventual climate effectiveness, this training should also apply to designers. This amplifies the message that evaluation and design cannot be treated as separate disciplines.

Training for improved evaluation and design

The approach used in most of the evaluations listed in Appendix 1 required a certain amount of on-the-job training for the teams involved. In large teams and with numerous investments to be evaluated, responsibilities had to be divided among team members, so it was important that all shared a common understanding of the criteria, the nature of relevant evidence, and the scoring system. It was found, however, that this could be achieved with a few hours' training, plus simultaneous trial assessments of the same intervention by the co-workers and team leader and discussion of the results (see Box 2.36). The response from team members to this imposition was wholly positive. The ease and reliability with which the techniques were transferred, and the appreciation of them, were encouraging, but raised the question of how best to promote a more unified approach among the many institutions and actors that comprise the aid world.

A book such as this can only go so far in conveying to aid professionals the principles and practices of systemic aid evaluation and design. Face-to-face courses will also be needed, to allow cases and examples to be explored and compared among participants in the context of their own professional development. A two-part course is suggested, the first focused on evaluation and design in general, and the second on the climate change response in particular.

- *Part 1: evaluation and design.* The first part would be modelled on Chapter 2, to explain and explore with examples how to use the core evaluation criteria of relevance, efficiency, effectiveness, impact, sustainability, connectedness, coherence, and replicability, and the other key considerations of partner satisfaction, cross-cutting themes, and design quality. Participants would learn why and how to use an extended version of the DAC evaluation criteria, with common definitions and scoring methods so that aid investments can be consistently understood, represented, and compared with each other and over time.
- *Part 2: the climate change response.* The second part would be modelled on Chapters 4–6, to practise applying the principles of evaluation and design to the various kinds of investment aimed at climate change mitigation and adaptation. Participants would learn how to analyse and interpret case material, while being sensitised to options and lessons learned, and how to deal confidently with the complexity of aid investments and the ecosystems and human societies where they occur. The need for this part of the course seems to be greatest in areas where mitigation and adaptation involve the management of ecosystems for services and values related to carbon, water, biological and cultural diversity, and resilience in general.
- *Parts 1 and 2: broader themes.* Both parts of the course would stress the need for understanding the historical, social, and ecological context of any investment as part of due diligence in evaluation or design. Informed by the content of Chapters 3 and 7–9, they would also build sensitivity to the issues surrounding partnerships, how to detect ripening circumstances and use them to induce transformations, and how to justify efforts directed to transformational aims. The course would also be a forum for people with diverse knowledge and experience, and an opportunity to encourage debate on such issues as the equitable distribution of wealth and opportunities in deteriorating environments, or the declining availability of paid work for people in increasingly machine-driven economic systems. These discussions might well yield valuable insights, as well as adding interest to the course for its participants.

Transforming aid and wider systems

The point of improving aid evaluation and design is to improve aid performance. The priority focus of direct investment is shifting and extending from poverty to climate change adaptation, and is joined by intense attention to enabling mitigation. Applying the principles and practices explained here across the aid portfolios of every donor, in concert with all their partners, would result in a steady increase in the performance of all their investments. Even this could only be a start, however, relative to the scale of the climate challenge. Climate sensitivity is needed not just in aid portfolios concerned with environmental sustainability, but in all aid investments of all kinds and in all sectors, and would have to include the resolution of strategic incoherence among policies and practices if net benefits are to be obtained. But even if all this was achieved, improving aid performance would still only affect directly a minute fraction of the world economy. Even with leverage and enhanced coherence, this would not be enough to make much difference to global outcomes unless there is very careful targeting of available resources.

For adaptation there will remain, for the foreseeable future, a vital role for aid in the following areas of care for diversity and resilience, all of them touched upon in Chapter 6:

- in research, education, knowledge exchange, networking, seed-grants and core funding for charitable activist groups, at all levels from the very local to the United Nations;
- in testing and demonstrating ideas and systems in such areas as cleaner production, disaster risk insurance, biodiversity inventory and bioprospecting, and payment for ecosystem services;
- in supporting and organising collective, often cross-border, services, in environmental monitoring, early warning, planning, conflict resolution, and policy development; and
- in filling persistent gaps in major areas of public investment, such as in protected areas and ecosystem restoration, disaster prevention and preparedness, local government capacity, and contingencies for humanitarian relief, reconstruction, and resettlement.

This adaptation list is already a long one, and may well grow longer, and fast, with improved performance being a high priority across all of it. If the proper role of aid is to induce the greatest benefit for the greatest number of the most vulnerable people, and everyone is vulnerable to climate change, then the adaptation agenda should be the target for the lion's share of all aid investment in the future.

The long-term role of aid is less obvious for mitigation, since what is needed is rapid change across whole economic systems, embracing trade, manufacture, tourism, and all other sectors, in which low-carbon development and negative-emission technologies are prioritised in all contexts and at all times. Here, the 'baby steps' described in this book are really just the beginning of a long journey. Like many past intellectual and physical movements that were forced upon peoples by changing circumstances, however, this would be a migration of hope and necessity as much as of choice. The early parts of the route are already mapped, and must involve the technologies, incentives, market operations, price signals, and negotiated arrangements reviewed in Chapters 4 and 5, but with a pressing need for creative problem solving, and brave and informed leadership.

Thus, the focus of mitigation aid should be to facilitate and enable structural, technological, and conceptual change, for instance through studies, prizes, and proof-of-concept grants, while reserving truly large public commitments for targets that can only be met publicly, or that can leverage fundamental changes in the direction and effects of private investments. This might, and should, urgently include the massive public investments that are needed to find, test and implement systems to pull GHGs out of the open air quickly and at gigatonne scale, and to store carbon safely, in functional isolation from the biosphere. Since these investments are required to prevent the destruction of the biosphere, they cannot really be seen as aid at all, but rather as a cost of doing business – like cleaning up any other industrial waste after profits have been taken. The fact that the nature of the waste was not appreciated (or was ignored) when the investments were originally made, is no reason for the beneficiaries not to meet the costs of clean up now, with those costs distributed in line with the benefits received in the past.

Setting aside – but not forgetting – the political challenge of mitigation, we should always be alert to ways for mitigation strategies to enhance the adaptation agenda, and the need to avoid strategic conflict between adaptation and mitigation investments. Where choices arise, the world can usually do without a short-term investment more than it can do without a natural ecosystem, a unique culture, or a just and sustainable social system, all of which represent the outcomes of long-term processes. It is important to remember this, since governments and commercial enterprises can deploy large amounts of capital, but were not designed to give priority to social, ecological, or long-term objectives beyond their immediate concerns, so tend to be careless of complex systems and permanent values. On the other hand, mitigation investments on the necessary scale may cause local or unforeseen impacts, and these will have to be learned from and compensated for as best we can. No doubt such choices will occupy human attention, and generate political anxieties, for ever.

Describing the kind of outcomes that are ultimately being sought can help to clarify the grand strategy of efforts and investments that aim to transform the human condition. In that spirit, part of a vision for the year 2085 is given in Box 9.3. Built into this account is the assumption that, in a world crowded with people sustained by damaged ecosystems, collective discipline is needed to ensure sustainability. This key point was made by Hardin (1993), who observed that where many people use the same ecosystem, they all have an incentive to over-exploit it. In other words, each participant has a motive to add one more demand and let the others bear the cost. The result is always disastrous – the so-called 'tragedy of the commons'. But the remedy is straightforward: all resource users must belong to a community that has agreed to be bound by rules on how to use the resource, and how to enforce compliance with those rules.

Hardin describes such an arrangement as '*mutual coercion, mutually agreed upon*', and offers it as the conceptual basis for all democratic governance. This is accepted into the definition of good governance that is used in this book: *stable, lawful, and effective governance maintained by accountability to an informed electorate.* Now consider that the biosphere is one ecosystem that is being used by all of us who participate in the global economic system. We were called 'biosphere people' by Dasmann (1998), to distinguish us from 'ecosystem people' who use only muscle power and very local resources to survive, and whose numbers and influence have been utterly eclipsed by us. In the big picture, therefore, putting Hardin's and Dasmann's insights together, we have a global ecosystem and a global population of people using it. Since a 'tragedy of the commons' is underway, the remedy must be a global community bound by rules on our own behaviour that comply with ecological reality. On this basis, the obvious strategic purpose of aid must be *to help build a functioning, inclusive, and ecologically lawful community among all people.* There are innumerable ways to contribute usefully to this aim, and high levels of performance are important to all of them.

Box 9.3 The year is 2085 ...

"On the High Seas, ships move under 24-hour satellite surveillance, their location continuously tracked and every movement in their fishing gear, cargo and ballast holds logged and measured. All merchant vessels are double hulled at least, and possess a host of other safety technologies developed during the '1.5-degree' storms that began in the first quarter of the century. The global fishing fleet is greatly reduced from early-21st-century levels, its vessels tightly licensed and closely supervised under agreements among every nation and federated region on Earth. Licence fees and freight taxes help pay for scientific studies and monitoring of all fish stocks and the health of marine ecosystems. The rules are amended from time to time in response to new data, with fine tuning through informed debate amongst the parties to global agreements. Policing is thorough and effective, punishment for cheating is harsh.

"Exclusive fishing areas and marine sanctuaries have been laid down by local councils along tens of thousands of kilometres of coast, safeguarding local resources and ensuring that any profits to be made are local ones. Enforcement in these areas is by local policing supported by national and federal government. Many communities are making full use of their property rights by joining biodiversity prospecting agreements with commercial firms and universities. These are searching for new products and processes among the immense array of evolved chemicals in living systems. Up-front payments and royalties from these agreements and discoveries have already transferred billions of dollars to networks of cooperating towns and villages.

"As a result of these new arrangements, and adaptation to climate change, fish stocks are showing signs of recovery over large areas of the sea, and permitted harvests are beginning to edge upwards for some species in several countries. This is starting to relieve prices in city shops, where wild-caught fish from the deep sea, like meat, has long been an expensive luxury. Most people's diets had not suffered from this, since in many areas near-shore fish have always been available from the management zones controlled by coastal communities, and there has generally been a good supply of farmed fish and vegetable food, albeit supplemented at times of famine by emergency rations. Now, however, the world's gourmets are looking forward to being able to buy some of the deep-sea fish that their old recipe books mentioned ...

"There are far fewer poisons around than there used to be. The agreement and enforcement of national laws and global treaties have seen to that, supported by the diligent activism of millions of citizens. Whistle-blowers, using mobile phone web connections to share direct observations with activists, journalists and enforcement agencies, make cheating almost impossible. The cost of safe waste disposal now has to be included in the annual business plans and accounts of all companies and cities, and the concept of 'safe disposal' has long since ceased to include dumping at sea. A result is that those sea fish that are available may be expensive, but they are at least safe to eat. Some businesses and towns went bankrupt when these laws were first put into effect, but environmental reconstruction grants and other forms of targeted public assistance smoothed the path for most. A clear legal framework for investment, and iron deadlines set well in advance, also helped the private sector make the necessary adjustments. Many companies were aided too by their customers, who deliberately chose only to buy the products of those who joined the new standards early on.

"But the most strategic change has been the stripping of greenhouse gases from the open air ... GHG concentrations in the atmosphere peaked at 0.05% carbon-dioxide equivalent in 2052, their growth having been slowed somewhat by earlier desperate measures to reduce emissions, but are now down to 0.04% and falling. The oceans stored a lot of energy in those few dangerous years in an enhanced greenhouse, and sea levels are still rising, rainfall patterns remain distorted, and wild storms still pound coastal areas. But there is a sense of hope nevertheless. Soft engineering, new building codes, and relocation of settlements and populations, has allowed for considerable adaptation, and most people are now reasonably safe.

"Tourism is still important, having recovered a bit from the almost-complete closure of commercial aviation during the mid-century famines. Now, high-speed air travel is used only for emergencies, weather permitting, and the usual way to travel long distances is by train. The new tourism is mostly a positive force for everyone. Coastal communities treat visitors rather like fishes, as a renewable resource for sustainable harvesting by locals. Almost every community in the world is in direct contact with every other through the web, and environmental education has been a big thing everywhere for decades. Thus standards are extremely high at the grass roots, and there is no question anywhere of shoddy or damaging tourism investments being allowed through ignorance of proper planning or equity standards. Corruption and incompetence have been minimized through direct local elections, the active participation of citizens' groups in government, and the free flow of information.

"Many resort areas that were ruined in earlier waves of tourism development, where they haven't had to be abandoned to the sea and sea-borne storms, have long since been rehabilitated, building on their unique strengths, the capacity of nature to regrow, and the friendliness of local people who no longer fear for their livelihoods. In any case, most visitors want to be ecotourists and are happy to pay what it takes for a high-quality holiday, without damaging the world too much by travelling around in it. The greenest of them, of course, stay home and tend their gardens instead. Or else they do voluntary ecosystem restoration work, driven by the need somehow to honour and atone for the deaths of tens of millions of species in the recent mass extinction.

"And what of fresh water in all this? The key change came with the mass movements of mid-century, in which the ideas of 'water democracy' became widespread. As locally accountable management of ecosystems became the norm, and communities learned from one another about what to require of their leaders, these ideas came to be expressed in a host of different ways, grafted onto a range of religions and philosophies of life ... In the same package came the strong sense of duty to use water sparingly, caringly and justly, and also the belief that all people and all species have a right to their necessary share of water. Many also concluded that water is unique and by nature a common resource, so it can't be owned as private property or sold as a commodity. In short, out of hardship, necessity, wisdom and local power, water became sacred again.

"The practical results were incredibly diverse, as people sought ways to solve their own local water crises, each resulting from a different local interaction among ecosystems, climate, weather, terrain, drainage, proximity to the sea, population, wealth and culture. And this was the whole point of local people seeking and gaining the power to make their own choices, putting their own water ethic into practice in their own way. They did this by restoring floodplains and welcoming floods, guiding rainwater down wells and into deep tanks, and building check-dams, all to recharge groundwater. They did it by harvesting dew, building sea-water greenhouses and other structures to condense water vapour from the air, using solar desalination, and catching and storing rain in domestic and community cisterns. They did it by banning thirsty crops, and seeking out species and cultivars that use least water. They did it by rediscovering ancient ways, such as underground tunnels, to harvest water from aquifers at sustainable rates. And they did it by finding old or new ways, such as bamboo or ceramic feeders, to deliver water drop by drop to the roots of growing crops, rather than smearing it across waterlogged and salinizing fields. All had their places in community water strategies around the world.

"Meanwhile, other configurations of cities, citizens, catchments, rivers and groundwaters came into play. City administrations learned to strike long-term deals with catchment dwellers to pay a fair price for catchment services, in return for their help in protecting upstream ecosystems. Cities learned to collaborate with one another to liberate rivers from industrial canals and dams, so that they and their shadows, and migrating fish, could run free again. They also learned to collaborate with land owners to encourage organic farming and low-impact use of ecosystems in the catchments. Their citizens, also affected by the ideas of 'water democracy', quickly became adept at conserving municipal waters, and began insisting that public water companies (the private ones having been repossessed) fix all leaks in their supply systems. They also began to demand that their urban environments become much more porous, with parks and gardens everywhere, all roads made of water-permeable materials, and a water trap on every roof. Finally, in a development that decisively changed the lives of consumers and producers around the world, sophisticated labelling and rationing arrangements for the use of virtual water came to be accepted, along with similar means to limit the use of embedded carbon."

Source: Caldecott (2008: 212–218).

References

Caldecott, J. O. (2008) *Water: the Causes, Costs and Future of a Global Crisis*. Virgin Books (London).

Caldecott, J. O., Halonen, M., Sørensen, S. E., Dugersuren, S., Tommila, P. & Pathan, A. (2010) *Evaluation of Sustainability in Poverty Reduction: Synthesis*. Evaluation report 2010: 4. Ministry for Foreign Affairs (Helsinki).

Caldecott, J. O., Dornau, R., Gollan, J., Halonen, M., Saalismaa, N., Schilli, A., Simonett, O., Stuhlberger, C. & Tommila, P. (2014) *Technical Report on Effectiveness of the Swiss International Cooperation in Climate Change Mitigation and Adaptation Interventions 2000-2012*. Mandate of the Swiss Agency for Development and Cooperation (SDC) and the State Secretariat for Economic Affairs (SECO) (Bern).

Dasmann, R. F. (1988) Towards a biosphere consciousness. Pages 177–188 in *The Ends of the Earth: Perspective on Modern Environmental History* (edited by D. Worster). Cambridge University Press (Cambridge).

Hardin, G. (1993) *Living Within Limits: Ecology, Economics and Population Taboos*. Oxford University Press (New York).

Appendix 1
Evaluations of aid performance used as main sources in this book

No.	Details
1	*Location:* Guyana. *Evaluation subject:* performance of the Iwokrama International Centre for Rainforest Conservation and Development and its strategic and business plans. *Evaluation criteria: sui generis* criteria for outputs related to: (1) institutional establishment; (2) sustainable forest management; (3) conservation and use of biodiversity; (4) sustainable human development; (5) public funding outlook; and (6) strategy for commercial development. *Methods used by the evaluation subject:* creation and operation of a national institution responsible for the sustainable management of 360,000 hectares of tropical moist forest as a proving ground for techniques of conservation and sustainable use. *Approach of the evaluation subject:* facilitating investment in creative use of the forest through reduced-impact logging, ecotourism, non-timber forest products, bioprospecting, carbon sequestration and trading, and endowments and grants, while minimising conflict and resource degradation, and ensuring that revenues are shared with the conservation and local human development processes in partnership with Amerindian communities. *Client(s):* EDG (Oxford), for DFID (London) on behalf of the Canadian International Development Agency, International Tropical Timber Organization & European Commission. *Reference:* Caldecott, J. O., Kowal, T. M., Holland, J. D. & Moran, D. (2001) *Report of the Multi-donor Mid-term Review of the Iwokrama International Centre for Rain Forest Conservation and Development, Guyana*. Department for International Development (London).
2	*Location:* Association of Southeast Asian Nations (ASEAN). *Evaluation subject:* performance of the EC-funded ASEAN Regional Centre for Biodiversity Conservation. *Evaluation criteria:* design quality, relevance, effectiveness, efficiency, impact, sustainability. *Methods used by the evaluation subject:* creation and operation of a regional institution for managing knowledge. *Approach of the evaluation subject:* reliance on regional cooperation to promote biodiversity conservation and climate change adaptation and mitigation. *Client(s):* IDL Group (Bristol) and GFA Terra Systems (Hamburg), for EU Delegation (Manila).

No.	Details
	Reference: Caldecott, J. O., Le Thac Can, Muhtaman, D. R., Scutt, B. & Tan, E. E. (2002) *Final Report of the Mid-Term Review of the ASEAN Regional Centre for Biodiversity Conservation*. EU Delegation (Manila).
3	*Location:* Indonesia. *Evaluation subject:* performance of the EC-funded Illegal Logging Response Centre. *Evaluation criteria:* relevance, effectiveness, efficiency, impact, sustainability. *Methods used by the evaluation subject:* creation and operation of a national institution for managing knowledge. *Approach of the evaluation subject:* reliance on institutional cooperation to suppress illegal logging and corruption through enhanced surveillance and case-tracking systems. *Client(s):* AGRIFOR (Les Isnes), for EU Delegation (Jakarta). *Reference:* Caldecott, J. O. & Teusan, S. (2006) *Final Report of the Pre-Completion Technical Evaluation Mission for the Illegal Logging Response Centre*. EU Delegation (Jakarta).
4	*Location:* Thailand. *Evaluation subject:* performance of the EC-funded Coastal Habitats and Resources Management (CHARM) Project. *Evaluation criteria:* relevance, effectiveness, efficiency, impact, sustainability. *Methods used by the evaluation subject:* building capacity for co-management at national, sub-national and community institutions. *Approach of the evaluation subject:* reliance on co-management of coastal resources by communities and government to demonstrate viable enterprises using living natural resources. *Client(s):* IDL Group (Bristol), for EU Delegation (Bangkok). *Reference*: Caldecott, J. O., Becker, A. & Udomittipong, P. (2007) *Final Evaluation of the Coastal Habitats and Resources Management (CHARM) Project*. EU Delegation (Bangkok).
5	*Location:* global. *Evaluation subject:* performance of policies and laws in solving environmental problems. *Evaluation criteria:* relevance, effectiveness, efficiency, impact, sustainability, connectedness, coherence, replicability. *Methods used by the evaluation subject:* national and sub-national policies and laws. *Approach of the evaluation subject:* reliance on policies and laws to reduce dependence on fossil fuels, to improve the management of urban, electronic and toxic wastes, to promote the use of payments for ecosystem services, to oppose desertification, over-fishing and biodiversity loss, to promote community-based forest management, ecosystem restoration and ecotourism, to safeguard cultural diversity, and to promote climate change adaptation. *Client(s):* CERTI (Colombo) and UN University (Yokohama), for UNEP (Nairobi). *Reference:* Caldecott, J., Chambers, B. & Kurukulasurya, L. (2008) *Green Breakthroughs: Solving Environmental Problems Through Innovative Policies and Law*. United Nations Environment Programme (Nairobi).

No.	*Details*
6	*Location:* Association of Southeast Asian Nations (ASEAN). *Evaluation subject:* performance of the EC-funded ASEAN Centre for Biodiversity. *Evaluation criteria:* design quality, relevance, effectiveness, efficiency, impact, sustainability. *Methods used by the evaluation subject:* creation and operation of a regional institution for managing knowledge. *Approach of the evaluation subject:* reliance on regional cooperation to promote biodiversity conservation and climate change adaptation and mitigation. *Client(s):* IDL Group (Bristol), for EU Delegation (Manila). *Reference:* Caldecott, J. O. & McNally, R. (2008) *Final Report of the Mid-Term Review of the ASEAN Centre For Biodiversity*. EU Delegation (Manila).
7	*Location:* global. *Evaluation subject:* performance of efforts to use ecosystems in promoting disaster risk reduction and climate change adaptation and mitigation. *Evaluation criteria:* effectiveness, efficiency, impact, sustainability, connectedness, replicability. *Methods used by the evaluation subject:* policies, projects and processes. *Approach of the evaluation subject:* reliance on ecosystem-based mitigation of earthquakes, landslides, avalanches, droughts and desertification processes, dry storms, fires, floods, cyclonic storms and tsunamis. *Client(s):* ProAct Network (Nyon), for the Ministry of the Environment of Finland (Helsinki). *Reference:* Caldecott, J. O. (2008) *The Role of Ecosystems in Disaster Risk Reduction and Climate Change Adaptation*. ProAct Network (Nyon).
8	*Location:* global. *Evaluation subject:* performance of the Multilateral Environmental Agreements (MEAs). *Evaluation criteria:* effectiveness, impact, sustainability, connectedness, coherence, replicability. *Methods used by the evaluation subject:* multilateral environmental treaties. *Approach of the evaluation subject:* reliance on the design, monitoring and enforcement of the MEAs on wetlands (Ramsar), endangered species (CITES Washington), migratory species (Bonn), biodiversity (CBD Rio, Cartagena), desertification (Paris), climate change (UNFCCC, Kyoto), ozone-depleting substances (Vienna, Montreal), hazardous wastes (Basel), hazardous chemicals in trade (Rotterdam) and persistent organic pollutants (Stockholm). *Client(s):* CERTI (Colombo), for UNEP (Nairobi). *Reference:* Caldecott, J. O. (2009) *Multilateral Environmental Agreements: A Global Response to the Crisis of Sustainability*. Centre for Environmental Research, Training and Information (Colombo).
9	*Location:* China. *Evaluation subject:* performance of the EC-funded EU–China Biodiversity Programme. *Evaluation criteria:* relevance, effectiveness, efficiency, impact, sustainability. *Methods used by the evaluation subject:* building capacity for biodiversity management at national, sub-national and community institutions. *Approach of the evaluation subject:* reliance on innovation to demonstrate that living natural resources can be used to sustain viable enterprises that contribute to biodiversity conservation, climate change adaptation and disaster risk reduction. *Client(s):* AGRIFOR (Les Isnes), for EU Delegation (Beijing).

No.	Details
	Reference: Caldecott, J. O., van Praet, C. & Cai, M. (2009) *Final Report of the Mid-Term Evaluation of the EU-China Biodiversity Programme*. EU Delegation (Beijing).
10	*Location:* Bolivia, India, Uganda (case studies). *Evaluation subject:* performance of the Finnish Partnership Agreement Scheme. *Evaluation criteria:* relevance, coherence, partner satisfaction, compatibility, Finnish added value, coordination, impact, effectiveness, sustainability, replicability, complementarity, efficiency, connectedness, activity design (synthesis evaluation). *Methods used by the evaluation subject:* outsourcing of complementary aid in the form of grants to non-governmental Partnership Organisations. *Approach of the evaluation subject:* reliance on capacity and motivations of PO grant recipients and their partners in developing countries. *Client(s):* Gaia Consulting Oy (Helsinki), for Ministry for Foreign Affairs of Finland (Helsinki). *Reference:* review of Virtanen, P., Mikkola, K. & Siltanen, M. (2008) *Finnish Partnership Agreement Scheme* (Evaluation report 2008: 1, Ministry for Foreign Affairs, Helsinki) in Caldecott, J. O., Halonen, M., Sørensen, S. E., Dugersuren, S., Tommila, P. & Pathan, A. (2010, Annex 6) *Evaluation of Sustainability in Poverty Reduction: Synthesis* (Evaluation report 2010: 4, Ministry for Foreign Affairs, Helsinki).
11	*Location:* Bolivia, India, Uganda (case studies). *Evaluation subject:* performance of the Finnish Local Cooperation Fund (LCF) system. *Evaluation criteria:* relevance, coherence, partner satisfaction, compatibility, Finnish added value, coordination, impact, effectiveness, sustainability, replicability, complementarity, efficiency, connectedness, activity design (synthesis evaluation). *Methods used by the evaluation subject:* non-governmental complementary aid in the form of embassy-managed grants. *Approach of the evaluation subject:* reliance on capacity and motivations of local civil-society applicants and implementers. *Client(s):* Gaia Consulting Oy (Helsinki), for Ministry for Foreign Affairs of Finland (Helsinki). *Reference:* review of Poutiainen, P., Mäkelä, M., Thurland, M. & Virtanen, P. (2008) *Evaluation of Local Cooperation Funds, Role in Institution Building of Civil Society Organizations* (Evaluation report 2008: 2, Ministry for Foreign Affairs, Helsinki) in Caldecott *et al.* (2010) – see evaluation number 10.
12	*Location:* Namibia. *Evaluation subject:* performance of a country programme. *Evaluation criteria:* relevance, coherence, partner satisfaction, compatibility, Finnish added value, coordination, impact, effectiveness, sustainability, replicability, complementarity, efficiency, connectedness, activity design (synthesis evaluation). *Methods used by the evaluation subject:* country-to-country aid programme using non-governmental complementary aid, institutional cooperation, concessional credits, economic and commercial partnerships, and regional mechanisms. *Approach of the evaluation subject:* reliance on programming and management to deliver effective development aid. *Client(s):* Gaia Consulting Oy (Helsinki), for Ministry for Foreign Affairs of Finland (Helsinki). *Reference:* review of Valjas, A., White, P., Thompson-Coon, R. & Gowaseb, K. (2008) *Evolving New Partnerships between Finland and Namibia* (Evaluation report 2008: 3, Ministry for Foreign Affairs, Helsinki) in Caldecott *et al.* (2010) – see evaluation number 10.

No.	*Details*
13	*Location:* global. *Evaluation subject:* performance of outsourcing arrangements to FIDIDA. *Evaluation criteria:* relevance, coherence, partner satisfaction, compatibility, Finnish added value, coordination, impact, effectiveness, sustainability, replicability, complementarity, efficiency, connectedness, activity design (synthesis evaluation). *Methods used by the evaluation subject:* outsourcing of disability services. *Approach of the evaluation subject:* reliance on capacity and motivations of FIDIDA to deliver and monitor disability services by their partners in developing countries. *Client(s):* Gaia Consulting Oy (Helsinki), for Ministry for Foreign Affairs of Finland (Helsinki). *Reference:* review of Hirstiö-Snellman, P. (2008) *FIDIDA: An Example of Outsourced Service 2004–2008* (Evaluation report 2008: 4, Ministry for Foreign Affairs, Helsinki) in Caldecott *et al.* (2010) – see evaluation number 10.
14	*Location:* global (132 countries). *Evaluation subject:* performance of three Finnish NGO foundations. *Evaluation criteria:* relevance, coherence, partner satisfaction, compatibility, Finnish added value, coordination, impact, effectiveness, sustainability, replicability, complementarity, efficiency, connectedness, activity design (synthesis evaluation). *Methods used by the evaluation subject:* outsourcing of complementary aid via Finnish NGO foundations that target disabilities (Abilis), human rights (KIOS), and environmental issues (Siemenpuu). *Approach of the evaluation subject:* reliance on capacity and motivations of Finnish foundations to deliver support to local civil society institutions in developing countries. *Client(s):* Gaia Consulting Oy (Helsinki), for Ministry for Foreign Affairs of Finland (Helsinki). *Reference:* review of Williams, P. J., Venäläinen, R. & Santisteban, R. (2008) *Finnish NGO Foundations* (Evaluation report 2008: 5, Ministry for Foreign Affairs, Helsinki) in Caldecott *et al.* (2010) – see evaluation number 10.
15	*Location:* Kosovo. *Evaluation subject:* performance of a country programme. *Evaluation criteria:* relevance, coherence, partner satisfaction, compatibility, Finnish added value, coordination, impact, effectiveness, sustainability, replicability, complementarity, efficiency, connectedness, activity design (synthesis evaluation). *Methods used by the evaluation subject:* country-to-country aid programme. *Approach of the evaluation subject:* reliance on programming and management to promote human rights and deliver basic services to vulnerable groups. *Client(s):* Gaia Consulting Oy (Helsinki), for Ministry for Foreign Affairs of Finland (Helsinki). *Reference:* review of Seppänen, M. & Karttunen, A. (2008) *Kosovo Country Programme* (Evaluation report 2008: 7, Ministry for Foreign Affairs, Helsinki) in Caldecott *et al.* (2010) – see evaluation number 10.
16	*Location:* Nepal, Nicaragua and Zambia (case studies). *Evaluation subject:* extent of mainstreaming of human rights, women's rights and gender equality, democracy, good governance and rule of law. *Evaluation criteria:* relevance, coherence, partner satisfaction, compatibility, Finnish added value, coordination, impact, effectiveness, sustainability, replicability, complementarity, efficiency, connectedness, activity design (synthesis evaluation).

No.	Details
	Methods used by the evaluation subject: linkage between policy, programming and management to deliver mainstreaming of cross-cutting themes (CCTs). *Approach of the evaluation subject:* reliance on mainstreaming of cross-cutting themes by embassies, project designers, etc. *Client(s):* Gaia Consulting Oy (Helsinki), for Ministry for Foreign Affairs of Finland (Helsinki). *Reference:* review of Kääriä, T., Poutiainen, P., Santisteban, R. & Pineda, C. (2008) *The Cross-cutting Themes in the Finnish Development* (Evaluation report 2008: 6, Ministry for Foreign Affairs, Helsinki) in Caldecott *et al.* (2010) – see evaluation number 10.
17	*Location:* Central Asia, South Caucasus, Eastern Europe. *Evaluation subject:* performance of a regional programme. *Evaluation criteria:* relevance, coherence, partner satisfaction, compatibility, Finnish added value, coordination, impact, effectiveness, sustainability, replicability, complementarity, efficiency, connectedness, activity design (synthesis evaluation). *Methods used by the evaluation subject:* regional programme. *Approach of the evaluation subject:* reliance on country-specific and regional projects, half delivered through multilateral channels, plus LCF grants. *Client(s):* Gaia Consulting Oy (Helsinki), for Ministry for Foreign Affairs of Finland (Helsinki). *Reference:* Review of Starr, S. F., Cornell, S. & Oksajärvi Snyder, M. (2009) *Finland's Development Cooperation in Central Asia and South Caucasus* (Evaluation report 2009: 1, Ministry for Foreign Affairs, Helsinki) in Caldecott *et al.* (2010) – see evaluation number 10.
18	*Location:* global. *Evaluation subject:* performance of development research. *Evaluation criteria:* relevance, coherence, partner satisfaction, compatibility, Finnish added value, coordination, impact, effectiveness, sustainability, replicability, complementarity, efficiency, connectedness, activity design (synthesis evaluation). *Methods used by the evaluation subject:* outsourcing of complementary aid to institutions for development-oriented research. *Approach of the evaluation subject:* reliance on grants to the Academy of Finland, directly commissioned, or as budget support to CGIAR, UN institutions, the World Bank and other organisations to promote research on the circumstances of developing countries and the international aid system, research cooperation with individuals and institutions in developing countries, and capacity building. *Client(s):* Gaia Consulting Oy (Helsinki), for Ministry for Foreign Affairs of Finland (Helsinki). *Reference:* review of Helland, J., Namaalwa, J. & Tostensen, A. (2009) *Support to Development Research* (Evaluation report 2009: 3, Ministry for Foreign Affairs, Helsinki) in Caldecott *et al.* (2010) – see evaluation number 10.
19	*Location:* Kenya. *Evaluation subject:* performance of an area-based programme. *Evaluation criteria:* relevance, coherence, partner satisfaction, compatibility, Finnish added value, coordination, impact, effectiveness, sustainability, replicability, complementarity, efficiency, connectedness, activity design (synthesis evaluation). *Methods used by the evaluation subject:* area-based country-to-country aid programme. *Approach of the evaluation subject:* reliance on consulting firms to deliver projects in rural water supply, health and livestock.

No.	Details
	Client(s): Gaia Consulting Oy (Helsinki), for Ministry for Foreign Affairs of Finland (Helsinki). *Reference:* review of Weir, A., Notley, M. & Katui-Katua, M. (2009) *Finnish Aid to Western Kenya* (Evaluation report 2009: 5, Ministry for Foreign Affairs, Helsinki) in Caldecott *et al.* (2010) – see evaluation number 10.
20	*Location:* global (23 countries). *Evaluation subject:* performance of support for cooperation among universities. *Evaluation criteria:* relevance, coherence, partner satisfaction, compatibility, Finnish added value, coordination, impact, effectiveness, sustainability, replicability, complementarity, efficiency, connectedness, activity design (synthesis evaluation). *Methods used by the evaluation subject:* outsourcing of complementary aid to cooperating universities. *Approach of the evaluation subject:* reliance on cooperation between Finnish and partner universities to build research and teaching capacity. *Client(s):* Gaia Consulting Oy (Helsinki), for Ministry for Foreign Affairs of Finland (Helsinki). *Reference:* review of Stenbäck, T. & Billany, N. (2009) *Evaluation of North-South-South Higher Education Institution Network Programme* (Evaluation report 2009: 7, Ministry for Foreign Affairs, Helsinki) in Caldecott *et al.* (2010) – see evaluation number 10.
21	*Location:* Nepal, Tanzania. *Evaluation subject:* performance of support for democratic political parties. *Evaluation criteria:* relevance, coherence, partner satisfaction, compatibility, Finnish added value, coordination, impact, effectiveness, sustainability, replicability, complementarity, efficiency, connectedness, activity design (synthesis evaluation). *Methods used by the evaluation subject:* outsourcing of complementary aid through DEMO Finland. *Approach of the evaluation subject:* reliance on grants to women's, students' and youth wings of political parties to promote transparency, networking activities, and open budget planning. *Client(s):* Gaia Consulting Oy (Helsinki), for Ministry for Foreign Affairs of Finland (Helsinki). *Reference:* review of Hällhag, R. & Sjöberg, F. M. (2009) *Political Parties for Democracy (DEMO Finland) Development Programme* (Evaluation report 2009: 6, Ministry for Foreign Affairs, Helsinki) in Caldecott *et al.* (2010) – see evaluation number 10.
22	*Location:* global (24 countries). *Evaluation subject:* performance of reporting systems for aid activities. *Evaluation criteria:* relevance, coherence, partner satisfaction, compatibility, Finnish added value, coordination, impact, effectiveness, sustainability, replicability, complementarity, efficiency, connectedness, activity design (synthesis evaluation). *Methods used by the evaluation subject:* meta-evaluation of reports on bilateral interventions, local cooperation funds, concessional credit, grants to international NGOs, contributions to international financial institutions, contributions to UN agencies, and partnerships with other donors in line with the 1997 Finnish development policy. *Approach of the evaluation subject:* reliance on high-quality reporting systems to promote learning. *Client(s):* Gaia Consulting Oy (Helsinki), for Ministry for Foreign Affairs of Finland (Helsinki).

No.	Details
	Reference: review of Williams, P. J. & Seppänen, M. (2009) *Meta-analysis of Development Evaluations in 2007 and 2008* (Evaluation report 2009: 9, Ministry for Foreign Affairs, Helsinki) in Caldecott *et al.* (2010) – see evaluation number 10.
23	*Location:* global. *Evaluation subject:* performance of reporting systems for HIV/AIDS-related activities. *Evaluation criteria:* relevance, coherence, partner satisfaction, compatibility, Finnish added value, coordination, impact, effectiveness, sustainability, replicability, complementarity, efficiency, connectedness, activity design (synthesis evaluation). *Methods used by the evaluation subject:* meta-evaluation of reports on ear-marked and core grants to bilateral and multilateral agencies. *Approach of the evaluation subject:* reliance on high-quality reporting systems to promote learning. *Client(s):* Gaia Consulting Oy (Helsinki), for Ministry for Foreign Affairs of Finland (Helsinki). *Reference:* review of Tuominen, M., Taylor, M. & Costa, D. (2009) *Meta-analysis of Development Cooperation on HIV/AIDS* (Evaluation report 2009: 4, Ministry for Foreign Affairs, Helsinki) in Caldecott *et al.* (2010) – see evaluation number 10.
24	*Location:* global. *Evaluation subject:* performance of Finnish support to the natural disasters and climate change sector. *Evaluation criteria:* relevance, coherence, partner satisfaction, compatibility, Finnish added value, coordination, impact, effectiveness, sustainability, replicability, complementarity, efficiency, connectedness, activity design (synthesis evaluation). *Methods used by the evaluation subject:* thematic projects and grants. *Approach of the evaluation subject:* reliance on capacity building to promote disaster preparedness and early warning. *Client(s):* Gaia Consulting Oy (Helsinki), for Ministry for Foreign Affairs of Finland (Helsinki). *Reference:* review of Srinivasan, G., Lehtonen, T., Munive, A. & Subbiah, A. (2009) *Natural Disasters and Climate Change in Finnish Aid from the Perspective of Poverty Reduction* (Evaluation report 2009: 8, Ministry for Foreign Affairs, Helsinki) in Caldecott *et al.* (2010) – see evaluation number 10.
25	*Location:* Zambia, Mozambique, Vietnam, Nicaragua, Kenya, Tanzania. *Evaluation subject:* performance of Finnish support to the agricultural sector. *Evaluation criteria:* relevance, coherence, partner satisfaction, compatibility, Finnish added value, coordination, impact, effectiveness, sustainability, replicability, complementarity, efficiency, connectedness, activity design (synthesis evaluation). *Methods used by the evaluation subject:* thematic projects and grants. *Approach of the evaluation subject:* reliance on projects and NGO activities to improve farming in poor areas. *Client(s):* Gaia Consulting Oy (Helsinki), for Ministry for Foreign Affairs of Finland (Helsinki). *Reference:* review of a draft of Ooijen, R., Lof, B., Wongtschowski, M. & Muchena, F. (2010) *Evaluation of Agriculture in the Finnish Development Cooperation* (unpublished, Ministry for Foreign Affairs, Helsinki) in Caldecott *et al.* (2010) – see evaluation number 10.

No.	Details
26	*Location:* Ethiopia, Nepal, Vietnam (case studies). *Evaluation subject:* performance of Finnish support to the water and sanitation sector. *Evaluation criteria:* relevance, coherence, partner satisfaction, compatibility, Finnish added value, coordination, impact, effectiveness, sustainability, replicability, complementarity, efficiency, connectedness, activity design (synthesis evaluation). *Methods used by the evaluation subject:* thematic projects and grants. *Approach of the evaluation subject:* reliance on projects, policies, implementation of the cross-cutting issues, and grant supplements to World Bank loans to improve water and sanitation. *Client(s):* Gaia Consulting Oy (Helsinki), for Ministry for Foreign Affairs of Finland (Helsinki). *Reference:* review of a draft of Matz, M., Blankwaardt, B. & Ibrahim-Huber, S. (2010) *Evaluation of Finnish Development Cooperation in the Water Sector* (unpublished, Ministry for Foreign Affairs, Helsinki) in Caldecott *et al.* (2010) – see evaluation number 10.
27	*Location:* Egypt. *Evaluation subject:* performance of a country programme. *Evaluation criteria:* relevance, coherence, partner satisfaction, compatibility, Finnish added value, coordination, impact, effectiveness, sustainability, replicability, complementarity, efficiency, connectedness, activity design (synthesis evaluation). *Methods used by the evaluation subject:* concessional credit, LCF, Finnpartnership, research exchange, institutional cooperation and higher education cooperation. *Approach of the evaluation subject:* reliance on projects, programming and management to deliver effective development aid on mental health, lands and surveys, antiquities, water and wastes. *Client(s):* Gaia Consulting Oy (Helsinki), for Ministry for Foreign Affairs of Finland (Helsinki). *Reference:* review of a draft of Wymenga, P., Sipilä. S., Zaki, S. S. & Tamminen, S. (2010) *Transition towards a New Partnership with Egypt* (unpublished, Ministry for Foreign Affairs, Helsinki) in Caldecott *et al.* (2010) – see evaluation number 10.
28	*Location:* Ethiopia. *Evaluation subject:* performance of a country programme. *Evaluation criteria:* relevance, coherence, partner satisfaction, compatibility, Finnish added value, coordination, impact, effectiveness, sustainability, replicability, complementarity, efficiency, connectedness, activity design (synthesis evaluation). *Methods used by the evaluation subject:* bilateral, multilateral and through Finnish NGOs, to Ethiopian civil society organisations, humanitarian assistance, and other instruments. *Approach of the evaluation subject:* reliance on projects, programming and management to deliver effective development aid on water, education, civil society and humanitarian assistance. *Client(s):* Gaia Consulting Oy (Helsinki), for Ministry for Foreign Affairs of Finland (Helsinki). *Reference:* review of a draft of MFA Finland (2010) *Finland's Development Cooperation with Ethiopia 2000–2008* (Ministry for Foreign Affairs, Helsinki) in Caldecott *et al.* (2010) – see evaluation number 10.

No.	*Details*
29	*Location:* Kenya, Laos PDR, Mozambique, Tanzania, Vietnam, Zambia, and regionally in Central America and Western Balkans. *Evaluation subject:* performance of Finnish support to the forestry and biological resources sector. *Evaluation criteria:* relevance, coherence, partner satisfaction, compatibility, Finnish added value, coordination, impact, effectiveness, sustainability, replicability, complementarity, efficiency, connectedness, activity design (synthesis evaluation). *Methods used by the evaluation subject:* thematic projects and grants. *Approach of the evaluation subject:* reliance on national forest programmes, international forest policy processes, strategic institutional partnerships, and projects concerned with the forest sector in rural development and climate change mitigation and adaptation. *Client(s):* Gaia Consulting Oy (Helsinki), for Ministry for Foreign Affairs of Finland (Helsinki). *Reference:* review of a draft of MFA Finland (2010) *Evaluation of Finnish Support to Forestry and Biological Resources* (Ministry for Foreign Affairs, Helsinki) in Caldecott *et al.* (2010) – see evaluation number 10.
30	*Location:* Kenya, Vietnam, Nicaragua, El Salvador. *Evaluation subject:* performance of Finnish support to the energy sector. *Evaluation criteria:* relevance, coherence, partner satisfaction, compatibility, Finnish added value, coordination, impact, effectiveness, sustainability, replicability, complementarity, efficiency, connectedness, activity design (synthesis evaluation). *Methods used by the evaluation subject:* thematic projects and contributions to bilateral, multi-bilateral and NGO projects. *Approach of the evaluation subject:* reliance on projects, programming and management to deliver: (a) improvements in energy pricing and energy security; (b) increased access to energy at national, regional and/or local scale; (c) improved energy efficiency and energy conservation technologies and practices; (d) increased renewable energy market share at national and/or regional scale; and (e) decreased GHG emissions from energy production. *Client(s):* Gaia Consulting Oy (Helsinki), for Ministry for Foreign Affairs of Finland (Helsinki). *Reference:* review of a draft of MFA Finland (2011) *Evaluation of the Sustainability Dimension in Addressing Poverty Reduction: Finnish Support to Energy Sector* (Evaluation report 2011: 1, Ministry for Foreign Affairs, Helsinki) in Caldecott *et al.* (2010) – see evaluation number 10.
31	*Location:* China, Vietnam (and elsewhere). *Evaluation subject:* performance of Finnish support to the concessional credit scheme. *Evaluation criteria:* relevance, coherence, partner satisfaction, compatibility, Finnish added value, coordination, impact, effectiveness, sustainability, replicability, complementarity, efficiency, connectedness, activity design (synthesis evaluation). *Methods used by the evaluation subject:* concessional credit scheme to finance Finnish exports to developing countries by relieving the buyer of interest payments. *Approach of the evaluation subject:* reliance on subsidised credit to promote Finnish commercial solutions to development issues. *Client(s):* Gaia Consulting Oy (Helsinki), for Ministry for Foreign Affairs of Finland (Helsinki). *Reference:* review of a draft of Weissenberg, T. von (editor, 2012) *Evaluation of Finnish Concessional Aid Instrument* (Evaluation report 2012: 4, Ministry for Foreign Affairs, Helsinki) in Caldecott *et al.* (2010) – see evaluation number 10.

No.	Details
32	*Location:* Indonesia. *Evaluation subject:* performance of the Indonesia–Norway REDD+ Partnership (first verification of deliverables). *Evaluation criteria: sui generis* criteria for outputs: (1) national REDD+ agency; (2) moratorium on forest and peatland concessions; (3) independent monitoring, reporting, and verification (MRV) institution; (4) funding instrument; (5) national REDD+ strategy; (6) work in pilot province(s); and (7) Indonesia–Norway Joint Consultation Group. *Methods used by the evaluation subject:* one billion US dollar commitment to finance verified deliverables in a process: (a) to build capacity to reduce GHG emissions from the land use, land-use change and forestry sector; and (b) to reward GHG emission reductions from avoided deforestation and peatland degradation. *Approach of the evaluation subject:* reliance on reform of the forest sector to enhance forest governance, transparency and reduced corruption, while establishing systems to allow MRV as a basis for payments for results. *Client(s):* Gaia Consulting Oy (Helsinki) and Creatura Ltd (Bath), for Government of Norway's International Climate and Forest Initiative (Oslo). *Reference:* Caldecott, J. O., Indrawan, M., Rinne, P. & Halonen, M. (2011) *Indonesia–Norway REDD+ Partnership: First Evaluation of Deliverables*. Government of Norway's International Climate and Forest Initiative (Oslo).
33	*Location:* Nepal. *Evaluation subject:* performance of a country programme during civil war, peace building, and peace time. *Evaluation criteria:* relevance, efficiency, strategic effectiveness, impact, sustainability, coordination, complementarity, compatibility, connectedness, coherence, Finnish added value, partner satisfaction, programming logic, replicability. *Methods used by the evaluation subject:* country-to-country aid programme comprising bilateral and multi-bilateral projects and contributions to multilateral projects, processes and institutions, with complementary LCF and NGO interventions and investment incentives. *Approach of the evaluation subject:* reliance on enabling peace and good governance, supporting the education sector, resource mapping, environmental administration and waste management, forest resources assessment and multi-stakeholder forestry, water and sanitation, environmental early warning, land administration, trade and investment partnerships. *Client(s):* Agrer (Brussels), for Ministry for Foreign Affairs of Finland (Helsinki). *Reference:* Caldecott, J. O., Hawkes, M., Bajracharya, B. & Lounela, A. (2012) *Evaluation of the Country Programme between Finland and Nepal*. Evaluation report 2012: 2. Ministry for Foreign Affairs (Helsinki).
34	*Location:* Nicaragua. *Evaluation subject:* performance of a country programme before, during and after a radical change in governance. *Evaluation criteria:* relevance, efficiency, strategic effectiveness, impact, sustainability, coordination, complementarity, compatibility, connectedness, coherence, Finnish added value, partner satisfaction, programming logic, replicability. *Methods used by the evaluation subject:* country-to-country aid programme comprising general budget support (to 2007), sector-wide approach programmes, bilateral and multi-bilateral projects, with complementary LCF and NGO interventions and regional initiatives on forestry, energy, and water and sanitation.

No.	Details
	Approach of the evaluation subject: reliance on rural development to reduce the vulnerability of the poor to economic exclusion, capacity building for primary and reproductive health care oriented to women's needs, and institutional capacity building for administrative effectiveness and decentralisation, with a later emphasis on good governance. *Client(s):* Agrer (Brussels), for Ministry for Foreign Affairs of Finland (Helsinki). *Reference:* Caldecott, J. O., van Sluijs, F., Aguilar, B. & Lounela, A. (2012) *Evaluation of the Country Programme between Finland and Nicaragua*. Evaluation report 2012: 1. Ministry for Foreign Affairs (Helsinki).
35	*Location:* Tanzania. *Evaluation subject:* performance of a country programme during Finnish policy and programming changes. *Evaluation criteria:* relevance, efficiency, strategic effectiveness, impact, sustainability, coordination, complementarity, compatibility, connectedness, coherence, Finnish added value, partner satisfaction, programming logic, replicability. *Methods used by the evaluation subject:* country-to-country aid programme comprising general budget support, bilateral and multi-bilateral projects, with complementary LCF and NGO interventions. *Approach of the evaluation subject:* reliance on projects in forestry, area-based rural development, and miscellaneous actions on telecommunications, energy, new institutional support, One UN pilot, refugees, seed potatoes, and geological mapping, plus civil society support. *Client(s):* Agrer (Brussels), for Ministry for Foreign Affairs of Finland (Helsinki). *Reference:* Caldecott, J. O., Valjas, A., Killian, B. & Lounela, A. (2012) *Evaluation of the Country Programme between Finland and Tanzania*. Evaluation report 2012: 3. Ministry for Foreign Affairs (Helsinki).
36	*Location:* Indonesia. *Evaluation subject:* performance of the Indonesia–Norway REDD+ Partnership (second verification of deliverables). *Evaluation criteria: sui generis* criteria for outputs: (1) national REDD+ agency; (2) moratorium on forest and peatland concessions; (3) independent monitoring, reporting, and verification (MRV) institution; (4) funding instrument; (5) national REDD+ strategy; (6) work in pilot province(s); and (7) Indonesia–Norway Joint Consultation Group. *Methods used by the evaluation subject:* one billion US dollar commitment to finance verified deliverables in a process: (a) to build capacity to reduce GHG emissions from the land use, land-use change and forestry sector; and (b) to reward GHG emission reductions from avoided deforestation and peatland degradation. *Approach of the evaluation subject:* reliance on reform of the forest sector to enhance forest governance, transparency and reduced corruption, while establishing systems to allow MRV as a basis for payments for results. *Client(s):* Gaia Consulting Oy (Helsinki) and Creatura Ltd (Bath), for Government of Norway's International Climate and Forest Initiative (Oslo). *Reference:* Caldecott, J. O., Rizki, D., Rinne, P. & Halonen, M. (2013) *Indonesia–Norway REDD+ Partnership: Second Evaluation of Deliverables*. Government of Norway's International Climate and Forest Initiative (Oslo).
37	*Location:* global. *Evaluation subject:* performance of Swiss efforts to mitigate climate change through renewable energy.

No.	Details
	Evaluation criteria: effectiveness. *Methods used by the evaluation subject:* projects, contributions to multilateral programmes, and government policy coordination. *Approach of the evaluation subject:* reliance on hydroelectric and biomass-based projects, earmarked contributions to the World Bank Renewable Energy Programme and Swiss government policy coordination. *Client(s):* Gaia Consulting Oy (Helsinki), Creatura Ltd (Bath), and Zoï Environment Network (Geneva), for Swiss Agency for Development and Cooperation & State Secretariat for Economic Affairs (Bern). *Reference:* Caldecott, J. O., Dornau, R., Gollan, J., Halonen, M., Saalismaa, N., Schilli, A., Simonett, O., Stuhlberger C. & Tommila, P. (2014) *Technical Report on Effectiveness of the Swiss International Cooperation in Climate Change Mitigation and Adaptation Interventions 2000–2012*. Mandate of the Swiss Agency for Development and Cooperation (SDC) and the State Secretariat for Economic Affairs (SECO) (Bern).
38	*Location:* global. *Evaluation subject:* performance of Swiss efforts to mitigate climate change through energy efficiency. *Evaluation criteria:* effectiveness. *Methods used by the evaluation subject:* projects and investment incentives. *Approach of the evaluation subject:* reliance on power system rehabilitation and sustainability, climate-friendly buildings and building materials, knowledge sharing among small enterprises, influencing consumer preferences, improving air quality, rehabilitating city-wide heating systems, and tram rehabilitation and re-use. *Client(s):* Gaia Consulting Oy (Helsinki), Creatura Ltd (Bath), and Zoï Environment Network (Geneva), for Swiss Agency for Development and Cooperation & State Secretariat for Economic Affairs (Bern). *Reference:* Caldecott *et al.* (2014) – see evaluation number 37.
39	*Location:* global. *Evaluation subject:* Performance of Swiss efforts to mitigate climate change through cleaner production. *Evaluation criteria:* effectiveness. *Methods used by the evaluation subject:* projects and investment incentives. *Approach of the evaluation subject:* reliance on national cleaner production centres, green credit trust funds and pollution abatement subsidies, earmarked contributions to the International Finance Corporation on environmental and social risk management, and recycling and other safe waste disposal (of CFCs and electronic wastes). *Client(s):* Gaia Consulting Oy (Helsinki), Creatura Ltd (Bath), and Zoï Environment Network (Geneva), for Swiss Agency for Development and Cooperation & State Secretariat for Economic Affairs (Bern). *Reference:* Caldecott *et al.* (2014) – see evaluation number 37.
40	*Location:* global. *Evaluation subject:* performance of Swiss efforts to mitigate climate change through ecosystem management. *Evaluation criteria:* effectiveness. *Methods used by the evaluation subject:* projects and investment/trade incentives. *Approach of the evaluation subject:* reliance on multi-stakeholder forest management, earmarked grants to multilateral institutions for landscape-level research,

No.	Details
	biotrade-based conservation, bamboo and forest plantation development and certification, dryland pasture ecosystem management, REDD+ with multiple stakeholders, and organic farming. *Client(s):* Gaia Consulting Oy (Helsinki), Creatura Ltd (Bath), and Zoï Environment Network (Geneva), for Swiss Agency for Development and Cooperation & State Secretariat for Economic Affairs (Bern). *Reference:* Caldecott *et al.* (2014) – see evaluation number 37.
41	*Location:* global. *Evaluation subject:* performance of Swiss efforts to promote climate change adaptation through risk management. *Evaluation criteria:* effectiveness. *Methods used by the evaluation subject:* projects and investment incentives. *Approach of the evaluation subject:* reliance on early warning systems for drought, flood, cold-weather disasters and glacial lake outbursts, capacity building and knowledge sharing, disaster risk insurance including inter-governmental risk sharing, micro-insurance for small-scale farmers and microcredit borrowers, and index systems to simplify and speed claims. *Client(s):* Gaia Consulting Oy (Helsinki), Creatura Ltd (Bath), and Zoï Environment Network (Geneva), for Swiss Agency for Development and Cooperation & State Secretariat for Economic Affairs (Bern). *Reference:* Caldecott *et al.* (2014) – see evaluation number 37.
42	*Location:* global. *Evaluation subject:* performance of Swiss efforts to promote climate change adaptation through farming systems and food security. *Evaluation criteria:* effectiveness. *Methods used by the evaluation subject:* projects. *Approach of the evaluation subject:* reliance on managing knowledge and improving regulation and networks to promote use of improved seed varieties and planting material for tree crops, and new ways to improve resilience and productivity of farming and food storage systems and soils and water-bearing ecosystems. *Client(s):* Gaia Consulting Oy (Helsinki), Creatura Ltd (Bath), and Zoï Environment Network (Geneva), for Swiss Agency for Development and Cooperation & State Secretariat for Economic Affairs (Bern). *Reference:* Caldecott *et al.* (2014) – see evaluation number 37.
43	*Location:* global. *Evaluation subject:* performance of Swiss efforts to promote climate change adaptation through water resources management (WRM). *Evaluation criteria:* effectiveness. *Methods used by the evaluation subject:* projects. *Approach of the evaluation subject:* reliance on area-based integrated WRM, knowledge management for WRM, payment for ecosystem services, physical and institutional rehabilitation of water systems, and water diplomacy and high-level capacity building. *Client(s):* Gaia Consulting Oy (Helsinki), Creatura Ltd (Bath), and Zoï Environment Network (Geneva), for Swiss Agency for Development and Cooperation & State Secretariat for Economic Affairs (Bern). *Reference:* Caldecott *et al.* (2014) – see evaluation number 37.

No.	Details
44	*Location:* global. *Evaluation subject:* performance of Swiss efforts to promote climate change adaptation through knowledge management. *Evaluation criteria:* effectiveness. *Methods used by the evaluation subject:* projects. *Approach of the evaluation subject:* reliance on environmental monitoring to improve the quality of more applied actions, such as planning to reduce risks and improve early warning, policy development through international knowledge sharing on technical, legislative and economic factors. *Client(s):* Gaia Consulting Oy (Helsinki), Creatura Ltd (Bath), and Zoï Environment Network (Geneva), for Swiss Agency for Development and Cooperation & State Secretariat for Economic Affairs (Bern). *Reference:* Caldecott *et al.* (2014) – see evaluation number 37.
45	*Location:* global. *Evaluation subject:* performance of Swiss efforts to promote climate change mitigation and adaptation through contributions to third-party agencies. *Evaluation criteria:* effectiveness. *Methods used by the evaluation subject:* non-earmarked contributions to multilateral institutions, thematic interest groups, NGOs and research institutions. *Approach of the evaluation subject:* reliance on enhanced effectiveness of beneficiaries from core donations or concessional loans, and/or policy or global synergies with thematic and site-specific interventions. *Client(s):* Gaia Consulting Oy (Helsinki), Creatura Ltd (Bath), and Zoï Environment Network (Geneva), for Swiss Agency for Development and Cooperation & State Secretariat for Economic Affairs (Bern). *Reference:* Caldecott *et al.* (2014) – see evaluation number 37.
46	*Location:* Vietnam. *Evaluation subject:* performance of Swiss efforts to promote climate change mitigation and adaptation through support to six projects in one country. *Evaluation criteria:* effectiveness. *Methods used by the evaluation subject:* support to two clusters of projects: on greening businesses (National Cleaner Production Centre, Green Credit Trust Fund, Environmental and Social Risk Management); and on biocommodities and air quality (forest-market linkages, bamboo investment conditions, and urban air quality management). *Approach of the evaluation subject:* Vietnam is an important partner for Swiss development cooperation, and the aim was to examine the effectiveness of the six individual projects as well as the extent of synergy among them over time. *Client(s):* Gaia Consulting Oy (Helsinki), Creatura Ltd (Bath), and Zoï Environment Network (Geneva), for Swiss Agency for Development and Cooperation & State Secretariat for Economic Affairs (Bern). *Reference:* Caldecott *et al.* (2014) – see evaluation number 37.
47	*Location:* Indonesia. *Evaluation subject:* performance of DFID's Forest Governance, Markets and Climate Programme. *Evaluation criteria: sui generis* output criteria: (1) on a multi-stakeholder deliberative process; (2) on forest sector governance reforms based on timber legality standards with civil society monitoring; (3) on land tenure reforms promoting land rights among forest-dependent peoples; and (4) on forest risk market

No.	Details
	reforms based on timber legality assurance and corporate behavioural change. *Methods used by the evaluation subject:* country-to-country aid programme. *Approach of the evaluation subject:* reliance on building capability for deliberation and for ensuring transparency and law enforcement, wide ownership of measures to support market-based changes in favour of sustainability and equity, and a systemic approach to systematically disabling the drivers of deforestation and forest degradation for multiple benefits and beneficiaries. *Client(s):* ITAD (Hove) and Triple Line Consulting (London), for DFID (London). *Reference:* Richardson, L., Caldecott, J. O. & Hakim, S. (2014) *Final Report of the Evaluation of DFID's Forest Governance, Markets and Climate Programme in Indonesia.* ITAD & Triple Line (London).
48	*Location:* Kenya. *Evaluation subject:* performance of a country programme before, during and after civil disorder and the introduction of a new constitutional settlement. *Evaluation criteria:* relevance, effectiveness, efficiency, impact, sustainability, the '3Cs' (complementarity, coherence and coordination), human rights/cross-cutting issues, connectedness. *Methods used by the evaluation subject:* country-to-country aid programme comprising bilateral, multi-bilateral and multilateral projects and contributions, with complementary LCF and NGO interventions. *Approach of the evaluation subject:* reliance on interventions on governance and gender, area-based agricultural development, dairy sector development, forest sector reform, and water and sanitation. *Client(s):* Agrer (Brussels), for Ministry for Foreign Affairs of Finland (Helsinki). *Reference:* Uusihakala, J., Caldecott, J., Killian, B. & Njagi, J. W. (2015) *Finland's Development Cooperation With Kenya in 2007–2013.* Evaluation report 2015: 2. Ministry for Foreign Affairs (Helsinki).
49	*Location:* global, and individual countries (Sudan and Sierra Leone). *Evaluation subject:* business cases for investments in fiscal systems, policing, knowledge management, and public health. *Evaluation criteria: sui generis* criteria for positive and negative climate and environment impacts, both by the intervention and on the intervention, for each feasible option related to each potential investment. *Methods used by the evaluation subject:* investment in improving performance of aid institutions (unilateral, or own-capacity focus), country-to-country aid programmes comprising bilateral, multi-bilateral and multilateral projects and contributions in multiple sectors. *Approach of the evaluation subject:* (a) building systems with which to identify and correct weaknesses in country fiscal systems. (b) Training and inter-institutional collaboration within the UK to building capacity among police intelligence and enforcement units against money laundering and corruption. (c) Improving the capacity across multiple donors and government in Sudan to collect and analyse high-quality data in support of donors' evidence-based decision making with a parallel contribution to a multi-donor trust fund for knowledge management, sector programmes and donor leverage. (d) Resourcing preventative and emergency services in the reproductive, maternal, newborn and child health sector in Sierra Leone. *Client(s):* Creatura Ltd (Bath), HTSPE Ltd (Hemel Hempstead) and IMC Worldwide Ltd (Redhill), for Department for International Development (London).

No.	Details
	References: Caldecott, J. O. (2014) *Climate and Environment Assessment: Support for the Tax Administration Diagnostic Assessment Tool.* (2014) *Climate and Environment Assessment: Tackling International Corruption in the UK through Anti-Money Laundering and Anti-Bribery Enforcement.* (2014) *Climate and Environment Assessment: Sudan Evidence Base Programme.* (2014) *Climate and Environment Assessment: Sudan Partnership Trust Fund.* (2015) *Climate and Environment Assessment: the 'Saving Lives, Improving Accountability, Strengthening Services' Project in the Reproductive, Maternal, Newborn and Child Health Sector in Sierra Leone.* Evidence on Demand, HTSPE Ltd (Hemel Hempstead).
50	*Location:* global (47 developing countries). *Evaluation subject:* performance of the World Bank-managed Forest Carbon Partnership Facility (FCPF). *Evaluation criteria:* relevance, efficiency, effectiveness, impact, sustainability (plus connectedness, coherence and replicability, and *sui generis* output criteria). *Methods used by the evaluation subject:* the FCPF aims to support developing countries in building their capacity to mitigate climate change by undertaking REDD+ (i.e. reducing GHG emissions from deforestation and forest degradation with the inclusion of environmental and social safeguards). *Approach of the evaluation subject:* preparing for REDD+ means developing appropriate policies, laws, strategies, institutions, maps, analyses, databases and monitoring systems. To do this the FCPF offers support from a Readiness Fund, and then from a Carbon Fund to help the countries define and implement their own REDD+ field activities and transactions. *Client(s):* Indufor Oy (Helsinki), for FCPF and the World Bank (Washington, DC). *Reference:* Caldecott, J., Robledo, C. & Clarke, M. (2015) *The World Bank: Second Program Evaluation of the Forest Carbon Partnership Facility (FCPF), Inception Report.* Indufor Oy (Helsinki) and Forest Carbon Partnership Facility (Washington, DC).

Appendix 2
Glossary, acronyms, and abbreviations

ACB. ASEAN Centre for Biodiversity.

Accra Platform (Accra Agenda for Action). A statement issued by the 2008 Accra High Level Forum on Aid Effectiveness, seeking to make the Paris Declaration more effective by improving: (a) *ownership* (countries should have more say over their own development through participation in policy formulation, leadership on aid coordination, and the use of country systems for aid delivery); (b) *inclusive partnerships* (all partners, including OECD/DAC members and other donors, developing countries, foundations and civil society, should participate more completely); (c) *delivering results* (aid should focus on achieving real and measurable impacts); and (d) *capacity development* (overall strategic priority should be given to building the ability of countries to manage their own futures).

Adaptation. The process by which an organism becomes better able to make a living in its environment. Also, a response by people to environmental change that makes life easier or more survivable than it would otherwise be.

Anthropocene. A name (from the Greek *anthrōpos* for 'human being', and *kainos* for 'new') proposed for the current (since around 1950) geological epoch to the International Commission on Stratigraphy by its Anthropocene Working Group (Waters *et al.*, 2016). The idea is that humanity is now having a profound influence on nature, including on "global atmospheric composition and temperature, cryosphere stability, ocean chemistry (e.g. acidification, anoxia) and sea level, biodiversity, landscape processes and near-surface sedimentation" (AWG, 2009: 1), and that this will be easily visible in the geological record all over the Earth for all future time.

Aquifer. A layer of water-bearing porous rock beneath the ground.

ARCBC. ASEAN Regional Centre for Biodiversity Conservation.

ASEAN. The Association of Southeast Asian Nations (comprising Brunei Darussalam, Cambodia, Indonesia, Lao People's Democratic Republic, Malaysia, Myanmar (Burma), the Philippines, Singapore, Thailand, and Vietnam).

Atmosphere. The envelope of gases around a planet. Earth's atmosphere has a well-mixed troposphere up to between 7 km (near the poles) and 17 km (near the equator), with a stratosphere above that to about 50 km, and other layers still higher. Close to the surface it contains water vapour, and when dry it is about 78 per cent nitrogen, 21 per cent oxygen, and 1 per cent argon, along with trace gases such as carbon dioxide.

Biodiversity. The variety and variation among everything alive, including ecosystems at all levels (from pond slime to the whole biosphere), species (which are temporarily stable aggregates of form and function within a sea of lineages all connected back to the origin of life on Earth), and genetic information (which codes for the forms and functions of all organisms).

Bioindicator. A group of organisms whose abundance or behaviour provides information on the condition of an ecosystem.

Biology. The study of how organisms live, reproduce, and evolve.

Biomass. The total weight of living material in a defined area.

Biomolecule. A molecule involved in the metabolic chemistry or structure of an organism.

Biopiracy. The acquisition of biodiversity information without first agreeing terms with the owners of that information.

Bioprospecting. The search for commercial products and processes using biodiversity as a source of information and as a stimulant for new ideas.

Biosphere. All parts of the Earth where life occurs, comprising the lower parts of the atmosphere, oceans, fresh waters, soils and land surfaces, including cities and farms, and their underlying sediments and shallow rock layers. Its total depth is about 30 km, from high altitude to depths underground that are too hot for organisms to survive.

Biota. All organisms found in a given area.

Biotechnology. Any skill, equipment, or procedure that is applied to organisms or ecosystems to make them more valuable to people.

CAMPFIRE. Communal Areas Management Programme for Indigenous Resources.

Carbon conservation (or sequestration). All actions that prevent carbon from being released to the air as a greenhouse gas, or that lead to it being recaptured and stored permanently out of contact with the air.

Carbon dioxide (CO_2). A greenhouse gas produced by most kinds of fire and metabolic activity.

Carbon dioxide equivalent (CO_2e). A measure of the potency of mixtures of greenhouse gases as agents of the greenhouse effect, expressed as a standard amount of CO_2. In tonnes (tCO_2e), this offers a standard unit for describing the significance of all emissions for global warming.

Catchment. An area where all water flows to a common destination, bounded by a line of terrain (the watershed) from which water can flow elsewhere.

CBD. Convention on Biological Diversity.

CBO. Community-based organisation.

CBRM. Community-based resource management.

CCTs. Cross-cutting themes, such as human rights, gender equality, good governance, and environmental sustainability, which are common to multiple sectors and pose special challenges for mainstreaming.

CERTI. Centre for Environmental Research, Training and Information.

CFCs. Chlorofluorocarbons. Chemicals widely used in refrigerators, air conditioners and industrial processes, but that when they escape destroy ozone (O_3) in the upper atmosphere, thereby weakening the ozone layer's ability to screen harmful radiation from sunlight. The CFCs are being phased out under the terms of the Vienna Convention and its Montreal Protocol, but it will be many decades before they are cleared from the atmosphere entirely. Meanwhile they also act as greenhouse gases.

CGIAR. Consultative Group on International Agricultural Research.

CIF. Climate Investment Funds.

CIFOR. Centre for International Forestry Research.

Clean Development Mechanism (CDM). An arrangement created by the Kyoto Protocol to the UNFCCC that lets a country earn emission-reduction credits through a project in a developing country.

Climate change. Defined in Article 1 of the UNFCCC as "a change of climate which is attributed directly or indirectly to human activity that alters the composition of the global atmosphere and which is in addition to natural climate variability observed over comparable time periods". This definition acknowledges the fact that climate change can occur naturally (e.g. from variation in solar output, or volcanic activity), but it is used here specifically to mean the more immediate and direct consequence of people having increased the amount of GHGs in the atmosphere, which has caused an enhanced greenhouse effect that is stressing the biosphere in many dangerous ways.

Co-evolution. The process by which one species comes to be dependent on another for fulfilling a vital need such as micro-nutrient supply, pollination or seed dispersal, so that neither can survive and/or reproduce in each other's absence.

Conservation. All actions that are based on and promote the idea that biodiversity and ecosystems are valuable.

CoP. Conference of the Parties (to an international convention).

CREMA. Community Resource Management Area (in Ghana).

CRIS. Common RELEX (External Relations) Information System (of the EC).

CSO. Civil-society organisation.

Cyclone. An area of low atmospheric pressure surrounded by inward-spiralling winds. Over warm seas, a cyclone picks up heat and moisture to become a powerful, rotating, moving storm. The strongest of these, with sustained wind speeds of over 120 km/hour, are known as *hurricanes*, *typhoons*, *tropical cyclones*, or *severe cyclonic storms*.

DAC. Development Assistance Committee (of the OECD).

Desertification. A process of land degradation that usually involves a combination of over-grazing and poor farming practices that expose the soil, drought that weakens it, and wind that blows it away.

DFID. Department for International Development (of the UK).

DRR. Disaster risk reduction.

EC. European Commission.

Ecology. The study of how organisms live together, meet their needs for energy and nutrients, and respond to opportunities and challenges in their environment.

Economics. The study of human preferences among all the various systems that we could invest in, among the various kinds of reward we might expect, and among the range of time intervals before we hope to receive them. These preferences are often explored using price indicators or monetary symbols.

Ecosystem. All the organisms living in a place and time, all the relationships among them, all the physical features of light, heat, moisture, wind, waves and chemistry that affect them, and the history of the place as well.

Ecosystem (or ecological) service. A function of an ecosystem that is of benefit to people, such as the storage of carbon by forests and coral reefs, the prevention of floods, water shortages, and soil erosion by vegetation on water catchments, the supply of fish and shellfish and prevention of coastal erosion by mangroves, and the digestion of sewage by wetlands.

Ecotourism. Nature-based tourism that uses but does not damage natural ecosystems and that contributes to their continued protection.

e.g. 'for example' (from the Latin *exempli gratia*).

EIA. Environmental impact assessment.

Elemental carbon (C). Carbon as a pure element, distinguished from all its compounds with other elements (e.g. carbon dioxide CO_2, methane CH_4).

El Niño–Southern Oscillation (ENSO). A global cycle that links oceans and the atmosphere, and is a major source of variation between years in rainfall around the world. El Niño events are marked by a switching in the intensity and direction of currents and winds in the Pacific Ocean, and are associated with droughts and floods in various locations across the world.

Endemic. A species or higher taxon, like a genus or a family, which occurs in the wild nowhere other than in a particular place, and which has not done so in the historical past.

Endemism rate. The percentage of species found nowhere other than the place for which the rate is calculated.

Equity. Fairness of treatment for all according to their needs, being either equal treatment, or treatment that is equivalent in terms of rights, benefits, obligations, and opportunities.

ESRM. Environmental and social risk management.

EU. European Union.

Ex-situ (or off-site) conservation. Maintaining organisms and biodiversity in artificial conditions (such as in a cage, gene-bank, or farm), rather than in the wild in a natural ecosystem.

Extinction. The death of the last member of a species, genus, family, etc.

FAO. Food and Agriculture Organization of the United Nations.

FCPF. Forest Carbon Partnership Facility.

FIP. Forest Investment Programme.

FLEGT. Forest Law Enforcement, Governance and Trade.

FONAFIFO. The Costa Rican Forest Financing Fund (*Fondo de Financiamiento Forestal*).

FSC. Forest Stewardship Council.

GCTF. Green credit trust fund.

GDP. Gross domestic product, the sum of all financial transactions in a place over a year.

GEF. Global Environment Facility.

Gene. Part of a DNA molecule that encodes for a single enzyme or structural protein molecule. Related to: *gene flow* (the mixing of genes between populations); *gene pool* (the genes within a freely-interbreeding population); *genetics* (the study of genes, lineages, genomes, evolution, speciation, genetic erosion, gene flow, and related subjects); *genetic erosion* (the reduced variability among members of a species due to reduction or loss of distinctive lineages); and *genome* (all genes possessed by an individual, a population, or a species).

Genus. A category of organisms at a taxonomic level just beneath that of the family and just above that of the species (i.e. a group of related, similar species). Plural: genera.

GESI. Gender equity and social inclusion.

GHG (greenhouse gas). A gas that is transparent to visible light but less so to heat energy, thus contributing to the greenhouse effect in the biosphere by trapping sunlight as heat. Important GHGs include water vapour (H_2O), carbon dioxide (CO_2), methane (CH_4), nitrous oxide (N_2O), sulphur hexafluoride (SF_6) and compounds based on bonds between atoms of carbon and fluorine (e.g. the chlorofluorocarbons or CFCs).

Global warming. An increase in average temperatures within the biosphere due to an enhanced greenhouse effect (i.e. the trapping on Earth of heat from the Sun) resulting from the accumulation of greenhouse gases in the atmosphere.

GLOF. Glacier lake outburst flood.

GNI. Gross national income (GDP adjusted for external debt payments).

Good governance. Stable, lawful, and effective governance maintained by accountability to an informed electorate.

Greenhouse effect. A process by which energy from the Sun is trapped. In a real greenhouse, light enters through the glass windows and is absorbed by and warms everything inside. Warm things radiate heat, which cannot pass through the glass, and the heat is therefore trapped inside, making it hotter. The atmosphere of the Earth acts rather like the glass in the greenhouse, and it blocks heat to the extent that it contains certain kinds and concentrations of greenhouse gas.

Habitat. The ecosystem inhabited by a population, species, or community.

Hectare (ha). An area of 100 × 100 metres, or 10,000 m^2.

IADB. Inter-American Development Bank.

ICDP. Integrated conservation-development project.

Ice age. A period when the world's climate cooled to cause a dramatic increase in the amount of ice accumulating in polar and mountain regions.

ICRAF. The World Agroforestry Centre (formerly the International Centre for Research in Agroforestry).

IDEA. Institute for Democracy and Electoral Assistance.

i.e. 'that is' (from the Latin *id est*).

IEG. Independent Evaluation Group (of the World Bank).

IFAD. International Fund for Agricultural Development.

IFC. International Finance Corporation (of the World Bank Group).

ILO. International Labour Organization (of the UN).

INBio. The Costa Rican National Biodiversity Institute (*Instituto Nacional de Biodiversidad*).

Indigenous peoples. Peoples who identify themselves as such, usually based on a combination of cultural distinctiveness and prior territorial occupancy relative to a more recently-arrived population that has its own distinct and often dominant culture.

In-situ (on-site) conservation. Maintaining organisms and biodiversity in the wild in a natural ecosystem.

IPCC. Intergovernmental Panel on Climate Change. Established by the WMO and UNEP in 1988, the IPCC organises the review and synthesis of information from published sources and governments on the risks of human-induced climate change, its potential impacts, and options for adaptation and mitigation.

ITTO. International Tropical Timber Organization.

IUCN. International Union for Conservation of Nature.

Kahriz. A tunnel system, also known as *qanat*, traditionally used to transport ground water from aquifers.

LCD. Low-carbon development.

LCF. Local Cooperation Fund (of Finland, managed through Finnish embassies).

LDC. Least-developed country.

Leverage. The power to use small assets to create large changes, for example by using know-how to amplify effects or replicate actions.

Lineage. A line of organisms descended over many generations, including kin, clans, cultivars, breeds, races, and sub-species.

LULUCF. Land use, land-use change and forestry, used as short-hand for a 'sector' that may also be called 'agriculture, forestry and other land use' (AFOLU).

MA. Millennium Ecosystem Assessment.

Macroscopic. Items or organisms that are large enough to be visible to the naked eye.

Mainstreaming. Bringing techniques and concerns associated with a particular specialised priority (such as biodiversity, climate change, or gender equity) into the standard procedures of an institution, so that it is thereafter routinely considered and addressed in all its work.

Mangrove. A type of tropical, coastal swamp forest made of trees that can grow in salty, tidally-flooded mud.

Mass extinction. A brief period when many species and lineages were lost almost simultaneously from the fossil record, events that are known from around 444 million, 350 million, 251 million, 200 million and 65 million years ago.

MDG. Millennium Development Goal. One of eight goals established by the United Nations in 2000, each with specific targets, to be achieved by 2015. The MDGs were: to eradicate extreme poverty and hunger; to achieve universal primary education; to promote gender equality and empower women; to reduce child mortality; to improve maternal health; to combat HIV/AIDS, malaria, and other diseases; to ensure environmental sustainability; and to develop a global partnership for development. In 2015, the MDGs were replaced by the SDGs.

Methane (CH_4). A potent greenhouse gas generated through metabolism and decay. The sudden release of large amounts of methane from melting Arctic sea-bed deposits and peat in permafrosts, is expected to cause a surge in global warming (Wadhams, 2016).

MFA. Ministry for Foreign Affairs (of Finland, Sweden, Norway, Denmark, and the Netherlands).

Microclimate. The conditions of shade, temperature, moisture, wind, etc., that prevail in a particular small location.

MoU. Memorandum of understanding.

MRV. Monitoring, reporting, and verifying emission reductions claimed for a REDD+ or other carbon conservation process.

NCPC. National cleaner production centre.

Necromass. Undecayed, dead material created and deposited by living or once-living ecosystems.

NETs. Negative-emission technologies.

NGO. Non-governmental organisation, usually a non-profit, public-interest group that is registered as a charity.

NICFI. Norway's International Climate and Forest Initiative.

NTFP. Non-timber forest product.

ODA. Official Development Assistance.

OECD. Organisation for Economic Co-operation and Development.

Organism. A living thing, which might be a plant, an animal, a fungus, or a microbe such as a bacterium, archaean, or protist.

PA. Protected area.

Paris Agreement. An international treaty made at the December 2015 UNFCCC Conference of the Parties to guide the world's climate change response, in terms of mitigation, adaptation, and finance, from the year 2020. It entered into force in November 2016, and includes a commitment to limit the increase in global mean temperature above pre-industrial levels to below 2°C during the 21st century.

Paris Declaration. A statement issued by the 2005 Paris High Level Forum on Aid Effectiveness, signed on behalf of donor and developing-country governments, multilateral donor agencies, regional development banks, and international agencies. It offered five principles for making aid more effective: (a) *ownership* (developing countries set their own strategies for poverty reduction, improve their institutions and tackle corruption); (b) *alignment* (donor countries align behind these principles and use local administrative and other systems for delivering aid); (c) *harmonisation* (donor countries coordinate among themselves, simplify procedures, and share information to avoid duplication); (d) *results* (both developing countries and donors shift their focus to the results of development investment, and results get measured); and (e) *mutual accountability* (donors and partners hold each other accountable for development results).

Partnership. An undertaking to do something that is based on free negotiation, informed consent, and agreement among equal stakeholders.

Peat (peatland). An accumulated mass of fallen leaves, dead moss, and other plant remains that have not decayed because they contain high concentrations of tannins or are perpetually waterlogged, or both.

PES. Payment for ecosystem services. Also known as payment for ecological or environmental services (e.g. in Spanish, *pago por servicios ambientales*, as used in Costa Rica).

Rainforest. Forest growing where total annual rainfall exceeds 2,500 mm, and where there are no months that regularly have less than 100 mm of rain.

RED. Reducing (greenhouse gas) emissions from deforestation.

REDD. Reducing (greenhouse gas) emissions from deforestation and (forest) degradation.

REDD+. REDD with the inclusion of environmental and social safeguards. Since the 2010 UNFCCC CoP, REDD+ has meant promoting more sustainable forest management, enhancing forest carbon stocks for example through natural regeneration and assisted natural regeneration or enrichment planting in natural forests, reducing emissions from deforestation and forest degradation, and conserving carbon stocks, with due attention to co-benefits linked to the safeguarding of poverty avoidance, indigenous peoples, good governance, biodiversity conservation, and ecosystem services.

Region. Either a part of a country (e.g. a state, province, county, or distinctive area such as a mountain range, plateau, plain, or delta) or a group of nearby countries (e.g. those comprising South-east Asia, Central America, or southern Africa). These may be specified as *sub-national* and *supra-national* regions respectively.

RNR. Renewable natural resource.

Salinisation. The accumulation of salts in the topsoil of irrigated farmland, caused by the evaporation of irrigation water that has previously soaked into the soil and dissolved salts in deeper levels.

SCF. Strategic Climate Fund.

SCP. Sustainable consumption and production.

SDC. Swiss Agency for Development and Cooperation of the Federal Department of Foreign Affairs (Bern, Switzerland).

SDG. Sustainable Development Goal. One of 17 goals adopted by the United Nations in 2015, each with specific targets, to be achieved by 2030. The SDGs are: to end poverty in all its forms everywhere; to end hunger, achieve food security and improved nutrition and promote sustainable agriculture; to ensure healthy lives and promote well-being for all at all ages; to ensure inclusive and equitable quality education and promote lifelong learning opportunities for all; to achieve gender equality and empower all women and girls; to ensure availability and sustainable management of water and sanitation for all; to ensure access to affordable, reliable, sustainable and modern energy for all; to promote sustained, inclusive and sustainable economic growth, full and productive employment and decent work for all; to build resilient infrastructure, promote inclusive and sustainable industrialisation and foster innovation; to reduce inequality within and among countries; to make cities and human settlements inclusive, safe, resilient and sustainable; to ensure sustainable consumption and production patterns; to take urgent action to combat climate change and its impacts; to conserve and sustainably use the oceans, seas and marine resources for sustainable development; to protect, restore and promote sustainable use of terrestrial ecosystems, sustainably manage forests, combat desertification, and halt and reverse land degradation and halt biodiversity loss; to promote peaceful and inclusive societies for sustainable development, provide access to justice for all and build effective, accountable and inclusive institutions at all levels; and to strengthen the means of implementation and revitalise the global partnership for sustainable development.

SECO. State Secretariat for Economic Affairs of the Federal Department of Economic Affairs, Education and Research (Bern, Switzerland).

SFDCC. Strategic Framework on Development and Climate Change (of the World Bank Group).

SFOE. Swiss Federal Office for Energy.

SIDS. Small-island developing state.

SME. Small and medium enterprise.

Species. A group of organisms capable of interbreeding freely with each other but not with members of other species, often identified using distinguishing physical characteristics.

Species richness. The total number of different species in an area.

Stakeholder. Someone with something to gain or lose in a potential dispute over resources.

Storm surge. The destructive arrival on land of a volume of moving water that has been raised by the low pressure within a cyclonic storm.

STRADA. A sub-national REDD+ strategy in Indonesia (from *Strategi Daerah*).

Stranded asset. An asset that has lost some or all of its value, or has become a liability, due to unanticipated or premature changes in *technology* (e.g. that render existing equipment or industrial processes obsolete), *regulation* (e.g. that prohibit forms of pollution that are integral to the asset concerned, including restrictions on carbon emissions that can make fossil fuel reserves unburnable and therefore valueless), *fashion* (e.g. where a company is holding large quantities of 'last year's model'), or *environmental context* (e.g. the failure of water supplies that are needed for a plantation or factory, due to climate change or damage to catchment ecosystems, or the loss of crop pollinators such as honey bees, due to misuse of insecticides).

Sub-species. A distinctive population that is not sufficiently different from other populations of the same species to be unable to interbreed with them if the opportunity arises.

Sui generis. A phrase in Latin, meaning 'of its own kind' or 'unique'.

Sustainability. From the Latin *sustinere*, from 'sub-' meaning 'from below' plus *tenere* meaning 'hold', implying something that is held up securely by an underlying structure, hence more likely to be found in 'bottom-up' than in 'top-down' arrangements of all kinds, and particularly in social and ecological ones. Related to: *sustainable* (something that can be maintained, and therefore, of an ecosystem, able to continue more-or-less unchanged despite being used by people); *sustainable development* (meeting the needs of current or local people in ways that preserve or enhance the ability of future or distant people to do the same); and *sustainable livelihood* (a means of living that does not undermine its own resource base, that can cope with and recover from stresses and shocks, and that maintains or enhances its value over time).

SVLK. The Indonesian Timber Legality Assurance System (*Sistem Verifikasi Legalitas Kayu*).

SWAp. Sector-wide approach programme.

Systemic evaluation. In ordinary English, 'systemic' means *relating to a system as a whole*, and 'system' means *a set of things working together as an interconnecting network*. The systemic evaluation approach uses multiple points of view to build up the fullest-possible appreciation of an aid investment as a complex, multi-faceted system in its own historical, social, and ecological context. Note the difference with 'systematic' which means *using a fixed plan for doing something*.

Taxon. A group of related organisms, such as a species, a genus (a group of related species), a family (a group of related genera) and an order (a group of related families). Plural: taxa.

Tenure. The holding by people of the means to meet their needs and control their futures, often based on agreements recognised by law or custom that define the rights and duties of ownership, access, and/or use of a particular resource.

ToR. Terms of reference.

UN. United Nations.

UNCCD. UN Convention to Combat Desertification in Countries Experiencing Serious Drought and/or Desertification, Particularly in Africa.

UNDP. UN Development Programme.

UNESCO. UN Educational, Scientific and Cultural Organization.

UNEP. UN Environment Programme.

UNEP-WCMC. UNEP's World Conservation Monitoring Centre.

UNFCCC. UN Framework Convention on Climate Change.

UNFPA. UN Population Fund (formerly the UN Fund for Population Activities).

Unicef. UN Children's Fund (formerly the UN International Children's Emergency Fund).

UNIDO. UN Industrial Development Organization.

WBG. World Bank Group.

WFP. World Food Programme.

WHO. World Health Organization.

WMO. World Meteorological Organization.

WRM. Water resources management.

WWF. World Wide Fund for Nature (also known as World Wildlife Fund).

References

AWG (2009) *Working Group on the 'Anthropocene' – Newsletter 1 (Dec 2009)*. Anthropocene Working Group of the Subcommission on Quaternary Stratigraphy (International Commission on Stratigraphy). http://quaternary.stratigraphy.org/workinggroups/anthropocene (consulted 8 Jan 2016).

Wadhams, P. (2016) *A Farewell to Ice*. Allen Lane (London).

Waters, C. N., Zalasiewicz, J., Summerhayes, C., Barnosky, A. D., Poirier, C., Gałuszka, A., Cearreta, A., Edgeworth, M., Ellis, E. C., Ellis, M., Jeandel, C., Leinfelder, R., McNeill, J. R., Richter, D. de B., Steffen, W., Syvitski, J., Vidas, D., Wagreich, M., Williams, M., An, Z.-S., Grinevald, J., Odada, E., Oreskes, N. & Wolfe, A. P. (2016) The Anthropocene is functionally and stratigraphically distinct from the Holocene. *Science*, **351**(6269): 138–141.

Index